PURE AND APPLIED MATHEMATICS

A Program of Monographs, Textbooks, and Lecture Notes

LECTURE NOTES
IN PURE AND APPLIED MATHEMATICS

1. *N. Jacobson,* Exceptional Lie Algebras
2. *L.-Å. Lindahl and F. Poulsen,* Thin Sets in Harmonic Analysis
3. *I. Satake,* Classification Theory of Semi-Simple Algebraic Groups
4. *F. Hirzebruch, W. D. Newmann, and S. S. Koh,* Differentiable Manifolds and Quadratic Forms
5. *I. Chavel,* Riemannian Symmetric Spaces of Rank One
6. *R. B. Burckel,* Characterization of C(X) Among Its Subalgebras
7. *B. R. McDonald, A. R. Magid, and K. C. Smith,* Ring Theory: Proceedings of the Oklahoma Conference
8. *Yum-Tong Siu,* Techniques of Extension of Analytic Objects
9. *S. R. Caradus, W. E. Pfaffenberger, and Bertram Yood,* Calkin Algebras and Algebras of Operators on Banach Spaces
10. *Emilio O. Roxin, Pan-Tai Liu, and Robert L. Sternberg,* Differential Games and Control Theory
11. *Morris Orzech and Charles Small,* The Brauer Group of Commutative Rings
12. *S. Thomeier,* Topology and Its Applications
13. *Jorge M. López and Kenneth A. Ross,* Sidon Sets
14. *W. W. Comfort and S. Negrepontis,* Continuous Pseudometrics
15. *Kelly McKennon and Jack M. Robertson,* Locally Convex Spaces
16. *M. Carmeli and S. Malin,* Representations of the Rotation and Lorentz Groups: An Introduction
17. *George B. Seligman,* Rational Methods in Lie Algebras
18. *Djairo Guedes de Figueiredo,* Functional Analysis: Proceedings of the Brazilian Mathematical Society Symposium
19. *Lamberto Cesari, Rangachary Kannan, and Jerry D. Schuur,* Nonlinear Functional Analysis of Differential Equations: Proceedings of the Michigan State University Conference
20. *Juan Jorge Schäffer,* Geometry of Spheres in Normed Spaces
21. *Kentaro Yano and Masahiro Kon,* Anti-Invariant Submanifolds
22. *Wolmer V. Vasconcelos,* The Rings of Dimension Two
23. *Richard E. Chandler,* Hausdorff Compactifications

Other volumes in preparation

Hausdorff Compactifications

RICHARD E. CHANDLER

Department of Mathematics
North Carolina State University
Raleigh, North Carolina

MARCEL DEKKER, INC. New York and Basel

MARCEL DEKKER, INC.

270 Madison Avenue, New York, New York 10016

LIBRARY OF CONGRESS CATALOG CARD NUMBER: 76-46693

ISBN: 0-8247-6559-1

Current printing (last digit):
10 9 8 7 6 5 4 3 2 1

PRINTED IN THE UNITED STATES OF AMERICA

For Suzanne

PREFACE

In 1930 Tychonoff discovered that those topological spaces which can be embedded in a compact Hausdorff space are precisely the completely regular (Hausdorff) spaces. This was essentially the beginning of the general study of Hausdorff compactifications since we can obtain a compactification of a space by embedding it in a compact space and then taking its closure. Once one compactification has been obtained, others can generally be constructed as quotient spaces of it. This, the typical means of construction in much of the literature on this subject, has one serious disadvantage from a technical point of view: it is frequently quite messy to show that the resulting quotient space is Hausdorff.

Tychonoff's original embedding was into a product of closed intervals, ingeniously using the set of bounded, continuous, real-valued functions as the indexing set of this product. By using an appropriate subset of this indexing set and proceeding in essentially the same way, any given Hausdorff compactification can be obtained. This will be the unifying theme here. We discuss other methods of construction and, when expedient, use them. However, most of the time we rely on variations on the original method. In some cases, a dramatic saving of effort is achieved: constructions in the original literature which took in the neighborhood of a journal page to effect are accomplished in a line or two.

These notes constitute a rewriting and enlargement of lectures given in a course at North Carolina State University during the Spring Semester of 1974. The only assumed prerequisite is an elementary course in general topology, of the type usually given to advanced undergraduate or beginning graduate students. In Chapter 1 we have attempted to bring students from the many possible such courses to a common viewpoint. We have attempted to be completely self-contained here with the exception of the elementary material alluded to above.

Each of the last four chapters ends with a statement of one of the major unsolved problems in the field which is appropriate to that chapter.

This should not be construed to imply that these are <u>the</u> problems of interest. Neither should the choice of subject matter be construed to imply that these are <u>the</u> topics of interest in the area. Both problems and topics represent personal choices.

I would like to express my appreciation to Dr. Robert Silber, Mr. Gary Brady, Mr. Eric Chandler, Mrs. Christine Lehmann, Mr. Michael Stadelmaier, Mr. Fu-Chien Tzung, and Mrs. Judy Williams for providing an attentive and enthusiastic audience for these lectures. I would also like to thank Mrs. Frances Jones for typing a preliminary version of the manuscript and Miss Gail Mattocks for typing the final version.

Richard E. Chandler

CONTENTS

Hausdorff Compactifications

§1. COMPLETELY REGULAR SPACES

The common separation properties usually involve distinguishing between pairs of points and/or closed sets with open sets. One exception to this is the property of complete regularity which uses continuous functions into the real numbers ($\mathbb{R}$) to separate points from closed sets not containing them. We thus need to consider certain families of such functions.

<u>1.1 Definitions.</u>

$C(X) = \{f\colon X \to \mathbb{R} \mid f \text{ is continuous}\}$

$C^*(X) = \{f\colon X \to \mathbb{R} \mid f \text{ is continuous and bounded}\}$

For functions in either of these sets we define

$(f + g)(x) \equiv f(x) + g(x)$

$(f \cdot g)(x) \equiv f(x) \cdot g(x)$

$(f \vee g)(x) \equiv \max\{f(x), g(x)\}$

$(f \wedge g)(x) \equiv \min\{f(x), g(x)\}$

$|f|(x) \equiv |f(x)|$

$Z(f) = \{x \in X \mid f(x) = 0\}$, the <u>zero set</u> of f.

For $r \in \mathbb{R}$, $\underline{r}(x) = r$ for all $x \in X$.

A Hausdorff space X is <u>completely regular</u> provided for each closed set $F \subset X$ and each $x \in X \setminus F$ there is an element $f \in C(X)$ such that $f(x) \notin \overline{f(F)}$.

<u>1.2 Remark</u>. This definition appears somewhat different from the usual one: there exists a continuous $g\colon X \to [0,1]$ such that $g(x) = 0$ and $g(F) = \{1\}$. They are equivalent.

<u>Proof</u>. We have $f(x) \notin \overline{f(F)}$ and need to find g. There exists $\delta > 0$ such that $(f(x) - \delta, f(x) + \delta) \cap f(F) = \phi$. Let $h\colon \mathbb{R} \to [0,1]$ be defined by $h(r) = (|r - f(x)|/\delta) \wedge \underline{1}$ and define $g = h \circ f$.

Clearly, the usual definition implies the condition in 1.1.

<u>1.3 Definition</u>. A collection $\mathcal{B}$ of subsets of a topological space X forms a <u>base for the closed sets</u> provided for each closed set $F \subset X$ and

each $x \in X \setminus F$ there is a $B \in \mathcal{B}$ with $F \subset B$ and $x \notin B$. A collection $\mathcal{S}$ of subsets of X forms a <u>subbase for the closed sets</u> provided the collection of all finite unions of elements from $\mathcal{S}$ forms a base for the closed sets.

<u>1.4 Lemma</u>. A Hausdorff space X is completely regular iff $\{Z(f) | f \in C^*(X)\}$ forms a base for the closed sets.

<u>Proof</u>. If X is completely regular then for a closed set F and a point $x \notin F$ there is $f \in C^*(X)$ such that $f(x) = 1$, $f(F) = \{0\}$. Then $F \subset Z(f)$ and $x \notin Z(f)$.

If $\{Z(f) | f \in C^*(X)\}$ forms a base for the closed sets then for any closed $F \subset X$ and any $x \notin F$ there is an $f \in C^*(X)$ for which $F \subset Z(f)$ and $x \notin Z(f)$. Clearly $f(x) \notin \overline{f(F)}$.

<u>1.5 Theorem</u>. A Hausdorff space X is completely regular iff its topology is the weakest for which each $f \in C^*(X)$ is continuous.

<u>Proof</u>. Suppose (X,τ) is completely regular, $\tau' \leq \tau$, and each $f \in C^*(X,\tau)$ is continuous with respect to τ'. If F is a closed set with respect to τ then for each $x \in X \setminus F$ there is an $f_x \in C^*(X)$ such that $f_x(x) = 1$ and $f_x(F) = \{0\}$. Since f_x is continuous with respect to τ', $Z(f_x)$ is closed with respect to τ'. Thus $F = \bigcap_{x \in X \setminus F} Z(f_x)$ is closed with respect to τ' and we see that $\tau' = \tau$.

Conversely, suppose τ is the weakest topology on X for which each $f \in C^*(X)$ is continuous. Then a subbase for the closed sets (with respect to τ) is the family

$\{\{x \in X | f(x) \geq r\} | f \in C^*(X), r \in \mathbb{R}\} \cup \{\{x \in X | f(x) \leq r\} | f \in C^*(X), r \in \mathbb{R}\}$.

We show that the base this family generates is the family of all zero sets of members of $C^*(X)$. The result then follows from 1.4.

First, every zero set $Z(f)$, $f \in C^*(X)$, is in this family since $Z(f) = \{x \in X | |f|(x) \leq 0\}$. A typical $\{x \in X | f(x) \geq r\} = Z(g)$ where $g = (f - \underline{r}) \wedge \underline{0}$ and $\{x \in X | f(x) \leq r\} = Z(h)$ where $h = (f - \underline{r}) \vee \underline{0}$. Now, a finite union of zero sets is a zero set: $Z(f_1) \cup \ldots \cup Z(f_n) = Z(f_1 \ldots f_n)$. It follows that the base generated by the family above is simply the family of all $Z(f)$, $f \in C^*(X)$.

<u>1.6 Corollary</u>. If X is a Hausdorff space whose topology is the weakest for which each $f \in C'$ is continuous (where C' is some family of bounded functions from X into $\mathbb{R}$) then X is completely regular.

Proof. Each $f \in C'$ is continuous so that $C' \subset C^*(X)$. Thus, the weak topology induced by C' is weaker than (or equal to) the weak topology induced by C^*. But the weak topology induced by C^* is obviously weaker than (or equal to) the topology on X, the weak topology induced by C'. These two topologies coincide and thus X is completely regular since the weak topology induced by C^* is completely regular (1.5).

We are now going to prove Urysohn's lemma which will imply that normal spaces are completely regular. The reason for needing this result here is that it provides the machinery for easily proving that compact Hausdorff spaces are completely regular since proving they are normal is routine.

1.7 Definition. A Hausdorff space X is normal if for each pair of disjoint closed sets A, B in X there is a pair of disjoint open sets U, V in X with $A \subset U$, $B \subset V$.

1.8 Theorem. [Urysohn [1925]]. If X is a normal space and A and B are disjoint closed sets in X then there is a continuous $f\colon X \to [0,1]$ with $f(A) = \{0\}$ and $f(B) = \{1\}$.

Proof. For each rational number r, we define an open set $U(r)$ in X as follows.

For $r < 0$, $U(r) = \phi$. For $r > 1$, $U(r) = X$. Enumerate the rationals in $[0,1]$ such that $r_1 = 1$, $r_2 = 0$ and let $U(r_1) = X \setminus B$. Since X is normal, there is an open set $U(r_2)$ containing A with $\overline{U(r_2)} \subset U(r_1)$. Suppose open sets $U(r_1), U(r_2), U(r_3), \ldots, U(r_{n-1})$ have been determined so that $\overline{U(r_i)} \subset U(r_j)$ if $r_i < r_j$. To determine $U(r_n)$, select r_k to be the largest of $r_1, r_2, \ldots, r_{n-1}$ which is smaller than r_n and select r_ℓ the smallest of $r_1, r_2, \ldots, r_{n-1}$ which is larger than r_n. By normality there is an open set $U(r_n)$ such that $\overline{U(r_k)} \subset U(r_n)$ and $\overline{U(r_n)} \subset U(r_\ell)$.

We have thus inductively defined the open sets $U(r)$ for each rational number r and $\overline{U(r)} \subset U(s)$ if $r < s$.

Define $f\colon X \to [0,1]$ by $f(x) = \inf\{r \mid x \in U(r)\}$. Since $\bigcup_r U(r) = X$ every $x \in X$ is in some $U(r)$. Also $\bigcap_r U(r) = \phi$ so no $x \in X$ is in all $U(r)$. Actually, $\{r \mid x \in U(r)\}$ is bounded below by 0 so that $f(x)$ is well-defined for all x. Since $x \in U(r)$ for all $r > 1$ it follows that $f(x) \leq 1$. Thus $f\colon X \to [0,1]$. It is easy to see that $f(A) = \{0\}$ and

$f(B) = \{1\}$. It remains to show that f is continuous.

The following facts are easily seen.

(i) $x \in U(r)$ implies $f(x) \leq r$.

(ii) $f(x) < r$ implies $x \in U(r)$.

(iii) $x \in \overline{U(r)}$ implies $x \in U(s)$ for all $s > r$, so that $f(x) \leq r$.

Fix $x_o \in X$ and let V be a neighborhood of $f(x_o)$ in $\mathbb{R}$. Choose rationals p and q such that $f(x_o) \in (p,q)$ and $[p,q] \subset V$. We claim that $U = U(q)\setminus\overline{U(p)}$ is a neighborhood of x_o and if $x \in U$ then $f(x) \in V$. First, $f(x_o) < q$ so that $x_o \in U(q)$ by (ii) and $f(x_o) \nleq p$ so $x_o \notin \overline{U(p)}$ by (iii). Thus, U is a neighborhood of x_o. Next, $x \in U$ implies $f(x) \leq q$ by (i) and $f(x) \geq p$ by (ii). It follows that f is continuous at x_o. Since x_o is an arbitrary point in X, we have f continuous on X.

<u>1.9 Corollary</u>. Normal spaces are completely regular.

<u>Proof</u>. Points are closed sets in a Hausdorff space.

<u>1.10 Definition</u>. A space X is <u>compact</u> if for every collection $\{0_\alpha\}_{\alpha\varepsilon A}$ of open sets with $X \subset \bigcup_{\alpha\varepsilon A} 0_\alpha$ there is a finite set of indices $\{\alpha_1,\ldots,\alpha_n\}$ with $X \subset 0_{\alpha 1} \cup\ldots\cup 0_{\alpha n}$. (This is usually stated by saying that "every open covering of X has a finite subcovering.") Contrapositively, this is equivalent to the condition that every family of closed sets with the finite intersection property (<u>i.e.</u>, every finite subfamily has a non-empty intersection) has a non-empty intersection.

<u>1.11 Theorem</u>. (i) Closed subsets of compact spaces are compact. (ii) Compact subsets of Hausdorff spaces are closed. (iii) If $f\colon X \to Y$ is continuous and X is compact, then $f(X)$ is compact. (iv) If $f\colon X \to Y$ is $1 - 1$ and continuous, X is compact, and Y is Hausdorff then f is a homeomorphism onto $f(X)$.

<u>Proof</u>. (i) Let $\{0_\alpha\}_{\alpha\varepsilon A}$ be an open cover of the closed subset F. Augment $\{0_\alpha\}_{\alpha\varepsilon A}$ with the open set $X\setminus F$ to obtain an open covering of the compact set X.

(ii) For a fixed point $x \in X\setminus F$ and each point y of F select disjoint open sets U_y, 0_y containing x,y. $\{0_y\}_{y\varepsilon F}$ is an open cover of F. Take a finite subcover and so the intersection of the corresponding U's will be a neighborhood of x in $X\setminus F$.

(iii) Let $\{O_\alpha\}_{\alpha\varepsilon A}$ be an open covering of $f(X)$. Then $\{f^{-1}(O_\alpha)\}_{\alpha\varepsilon A}$ is an open covering of X. The O_α corresponding to the finite subcovering of X provide a finite subcovering of $f(X)$.

(iv) For an open set $O \subset X$ we have $X\setminus O$ is closed; hence compact; hence $f(X\setminus O)$ is compact; hence $f(X\setminus O)$ is closed; hence $f(O)$ is open. Thus, $f: X \to f(X) \subset Y$ is an open mapping.

<u>1.12 Theorem</u>. If X is a compact Hausdorff space, X is normal.

<u>Proof</u>. Let A and B be disjoint closed subsets of X.

For a fixed $x \varepsilon A$ and each $y \varepsilon B$ there are disjoint open sets U_y, V_y containing x,y. $\{V_y\}_{y\varepsilon B}$ is an open covering of B, a compact set. Let V_x be the union of the elements of a finite subcovering and let U_x be the intersection of the corresponding U_y's. Then $U_x \cap V_x = \phi$, $x \varepsilon U_x$, and $B \subset V_x$. Repeat this process for each $x \varepsilon A$. Then $\{U_x\}_{x\varepsilon A}$ is an open covering of A, a compact set. Let U be the union of the members of a finite subcovering and let V be the intersection of the corresponding V_x's. Then $U \cap V = \phi$, $A \subset U$, $B \subset V$.

<u>1.13 Corollary</u>. Compact Hausdorff spaces are completely regular.

<u>Proof</u>. 1.12 and 1.9.

<u>1.14 Theorem</u>. For a completely regular space X the following are equivalent.

(i) X is compact

(ii) Every open covering of X with elements of a specific base has a finite subcovering.

(iii) Every family of closed sets from a specific base having the finite intersection property has a non-empty intersection.

(iv) For each maximal ideal $M \subset C^*(X)$ there is an $x \varepsilon X$ with $x \varepsilon \bigcap_{f\varepsilon M} Z(f)$. (In this case we say M is <u>fixed</u>.)

<u>Proof</u>. <u>(i) $\Leftrightarrow$ (ii)</u> Trivially (i) $\Rightarrow$ (ii). Conversely, let $\{O_\alpha\}_{\alpha\varepsilon A}$ be an open covering of X. Let $\mathcal{B}$ be the base for which open coverings from it have finite subcoverings and for each O_α and each $x \varepsilon O_\alpha$ choose an element from $\mathcal{B}$, say $V(x,\alpha)$ containing x and contained in O_α. Then $\{V(x,\alpha) \mid \alpha\varepsilon A \text{ and } x \varepsilon O_\alpha\}$ is an open covering for X. The O_α's corresponding to the finite subcovering of V's is a finite cover of X.
<u>(i) $\Leftrightarrow$ (iii)</u> Similar to above, using contrapositive statements.

<u>(i) ⇒ (iv)</u>. First observe that (i) implies every $f \in C(X)$ is bounded: otherwise $\{f^{-1}[(-n,n)] \mid n \in Z^+\}$ would be an open covering with no finite subcovering. This implies that an element of $C^*(X)$ is a unit (*i.e.*, invertible) if and only if $Z(f) = \phi$. Since M is an ideal it follows that $Z(f) \neq \phi$ for all $f \in M$. For $f_1,\ldots,f_n \in M$ we have $\bigcap_{i=1}^{n} Z(f_i) = Z(f_1^2 +\ldots+ f_n^2) \neq \phi$ $(f_1^2 +\ldots+ f_n^2 \in M)$ so it follows that $\{Z(f) \mid f \in M\}$ is a family of closed sets in X with the finite intersection property. By (i) there is a point $x \in X$ with $x \in \bigcap_{f \in M} Z(f)$.

<u>(iv) ⇒ (iii)</u>. We will show that (iii) holds for a specific base, the zero sets of elements from $C^*(X)$. This implies (i) holds which then implies (iii) holds for any other base.

Let $\{Z(f)\}_{f \in C'}$ be a family of zero sets with the finite intersection property. Let $I = \{g_1 f_1 +\ldots+ g_k f_k \mid k \in Z^+,\ g_i \in C^*(X),\ f_i \in C'\}$.

I is clearly an ideal in $C^*(X)$. Since $\underline{1} \notin I$, I is a proper ideal. (If $\underline{1} = g_1 f_1 +\ldots+ g_k f_k$ then for no $x \in X$ would it be true that $g_1(x)f_1(x) +\ldots+ g_k(x)f_k(x) = 0$. However $Z(f_1) \cap\ldots\cap Z(f_k) \neq \phi$.)

Let M be a maximal ideal in $C^*(X)$ containing I. Then

$$\phi \neq \bigcap_{f \in M} Z(f) \subset \bigcap_{f \in I} Z(f) \subset \bigcap_{f \in C'} Z(f).$$

<u>1.15 Remark</u>. It is apparent that the above proof uses the axiom of choice: most obviously in claiming the existence of a maximal ideal containing a given proper ideal. The axiom of choice is inextricably bound to much of topology involved with compactness. Of the many possible sources of information on this interrelationship, Comfort [1968] is very complete and does not require a great amount of technical knowledge of set theory to understand.

We see another example of this involvement in the next theorem we prove, Tychonoff's theorem on the compactness of the product of compact spaces. This was shown to be equivalent to the axiom of choice by Kelley [1950]. In proving this theorem we rely on another result equivalent to the axiom of choice. Many modern texts on set theory contain a proof of this equivalence, e.g. Jech [1973].

We first provide some relevant definitions concerning product spaces.

<u>1.16 Definitions</u>. Let $\{X_\alpha\}_{\alpha \in A}$ be a collection of topological spaces. Their <u>product</u>, $\prod_{\alpha \in A} X_\alpha$, is the set $\{f\colon A \to \bigcup_{\alpha \in A} X_\alpha \mid f(\alpha) \in X_\alpha\}$ with

topology generated by the subbase consisting of all sets of the form $\pi_\alpha^{-1}(O_\alpha)$, where O_α is open in X_α, and $\pi_\alpha: \prod X_\alpha \to X_\alpha$, the <u>projection map</u>, is defined by $\pi_\alpha(f) = f(\alpha)$. If $A = \{1,2,\ldots,n\}$ is a finite set, we denote $\prod_{\alpha\in A} X_\alpha$ by $X_1 \times X_2 \times\ldots\times X_n$.

A property of sets is <u>of finite character</u> if the empty set has it and a set F has it if and only if every finite subset of F has it.

<u>1.17 Lemma [Tukey [1940]]</u>. If $\mathcal{P}$ is a property of finite character which subsets of a set T may have, then a subset $T_o \subset T$ which has property $\mathcal{P}$ is contained in a maximal (with respect to inclusion) subset T_m having $\mathcal{P}$.

<u>1.18 Theorem [Tychonoff [1930]]</u>. If each X_α is compact, $\alpha \in A$, then so is $\prod_{\alpha\in A} X_\alpha$.

<u>Proof</u>. Let $\mathcal{F}$ be a family of closed subsets of $\prod_{\alpha\in A} X_\alpha$ having the finite intersection property. As this property is clearly of finite character, $\mathcal{F}$ is contained in a maximal family $\mathcal{F}_o$ (of subsets of $\prod_{\alpha\in A} X_\alpha$) having the finite intersection property.

Since $\mathcal{F}_o$ is maximal, we see that whenever $F_1,\ldots,F_k \in \mathcal{F}_o$ then $F_1 \cap\ldots\cap F_k \in \mathcal{F}_o$. Also, if $F \cap F_o \neq \emptyset$ for all $F \in \mathcal{F}_o$ then $F_o \in \mathcal{F}_o$.

Now for each $\alpha \in A$, $\{\overline{\pi_\alpha(F)}\}_{F\in\mathcal{F}_o}$ is a family of closed sets in X_α having the finite intersection property. Since X_α is compact, there is a point $x_\alpha \in \bigcap_{F\in\mathcal{F}_o} \overline{\pi_\alpha(F)}$. Let $x \in \prod_{\alpha\in A} X_\alpha$ be the element for which $x(\alpha) = x_\alpha$ for each α.

For each α, if U_α is an open set in X_α containing x_α, then $U_\alpha \cap \pi_\alpha(F) \neq \emptyset$ for each $F \in \mathcal{F}_o$ (since $x_\alpha \in \overline{\pi_\alpha(F)}$). Therefore, $\pi_\alpha^{-1}(U_\alpha) \cap F \neq \emptyset$ for each $F \in \mathcal{F}_o$. It follows that $\pi_\alpha^{-1}(U_\alpha) \in \mathcal{F}_o$. This holds for arbitrary open sets U_α in X_α containing x_α. We see that for any finite collection $\{\alpha_1,\ldots,\alpha_n\}$, $\bigcap_{i=1}^{n} \pi_{\alpha_i}^{-1}(U_{\alpha_i}) \in \mathcal{F}_o$. Thus $\mathcal{F}_o$ contains all basic neighborhoods of x in $\prod_{\alpha\in A} X_\alpha$ and so contains all neighborhoods of x in $\prod_{\alpha\in A} X_\alpha$. We conclude that for any $F \in \mathcal{F}_o$ and any neighborhood U of x in $\prod_{\alpha\in A} X_\alpha$ we have $U \in \mathcal{F}_o$ and therefore $U \cap F \neq \emptyset$. It follows that $x \in \overline{F}$ for all $F \in \mathcal{F}_o$. Thus,

$$\emptyset \neq \bigcap_{F\in\mathcal{F}_o} \overline{F} \subset \bigcap_{F\in\mathcal{F}} \overline{F} = \bigcap_{F\in\mathcal{F}} F.$$

As we will be using product spaces extensively we need some other basic results.

1.19 Proposition. $f\colon X \to \prod_{\alpha\in A} X_\alpha$ is continuous if and only if $\pi_\alpha \circ f\colon X \to X_\alpha$ **is continuous for each $\alpha \in A$.**

Proof. If f is continuous, so is $\pi_\alpha \circ f$.

Conversely, suppose $\pi_\alpha \circ f$ is continuous and let $\pi_\alpha^{-1}(O_\alpha)$ be a subbasic open set in $\prod_{\alpha\in A} X_\alpha$. Then $f^{-1}(\pi_\alpha^{-1}(O_\alpha)) = (\pi_\alpha \circ f)^{-1}(O_\alpha)$ which is open.

1.20 Definition. Let $A' \subset A$. We define the projection $\pi_{A'}\colon \prod_{\alpha\in A} X_\alpha \to \prod_{\alpha\in A'} X_\alpha$ by

$$[\pi_{A'}(f)](\alpha) = f(\alpha) \quad \text{for each } \alpha \in A'.$$

1.21 Proposition. $\pi_{A'}$ is a continuous, open mapping.

Proof. $\pi_\alpha \circ \pi_{A'} = \pi_\alpha$ for each $\alpha \in A'$ (where the π_α on the left side is $\pi_\alpha\colon \prod_{\alpha\in A'} X_\alpha \to X_\alpha$ and the π_α on the right side is $\pi_\alpha\colon \prod_{\alpha\in A} X_\alpha \to X_\alpha$.) Thus, by 1.19 $\pi_{A'}$ is continuous.

We show next that $\pi_{A'}$ takes basic open sets in $\prod_{\alpha\in A} X_\alpha$ to open sets in $\prod_{\alpha\in A'} X_\alpha$: First, note that if $\alpha \notin A'$ then

$$\pi_{A'}(\pi_\alpha^{-1}(O_\alpha)) = \prod_{\alpha\in A'} X_\alpha.$$

Thus, $\pi_{A'}(\pi_{\alpha_1}^{-1}(O_{\alpha_1}) \cap \ldots \cap \pi_{\alpha_n}^{-1}(O_{\alpha_n})) = \pi_{\alpha_1}^{-1}(O_{\alpha_1}) \cap \ldots \cap \pi_{\alpha_k}^{-1}(O_{\alpha_k})$ where we assume that $\{\alpha_1,\ldots,\alpha_n\}$ has been named in such a way that $\{\alpha_1,\ldots,\alpha_k\} = \{\alpha_1,\ldots,\alpha_n\} \cap A'$. (If $\{\alpha_1,\ldots,\alpha_n\} \cap A' = \emptyset$ then $\pi_{A'}(\pi_{\alpha_1}^{-1}(O_{\alpha_1}) \cap \ldots \cap \pi_{\alpha_n}^{-1}(O_{\alpha_n})) = \prod_{\alpha\in A'} X_\alpha$.) Thus, $\pi_{A'}$ takes basic open sets to open sets. Since functions preserve unions, $\pi_{A'}$ takes open sets to open sets.

1.22 Proposition. If each X_α is completely regular, so is $\prod_{\alpha\in A} X_\alpha$.

Proof. To show that $\prod_{\alpha\in A} X_\alpha$ is Hausdorff is routine.

Now if F is a subbasic closed set in $\prod_{\alpha\in A} X_\alpha$ then $F = \prod_{\alpha\in A} X_\alpha \setminus \pi_\alpha^{-1}(O_\alpha)$ for some $\alpha \in A$ and O_α open in X_α. If $x \in \pi_\alpha^{-1}(O_\alpha)$ there is a function $f_\alpha\colon X_\alpha \to \mathbb{R}$ with $f_\alpha(x(\alpha)) \notin \overline{f(X_\alpha \setminus O_\alpha)}$. Thus $f_\alpha \circ \pi_\alpha\colon \prod_{\alpha\in A} X_\alpha \to \mathbb{R}$ and $f_\alpha \circ \pi_\alpha(x) \notin \overline{f_\alpha \circ \pi_\alpha(F)}$.

<u>1.23 Definition</u>. Let $\{f_\alpha\colon X \to Y_\alpha\}_{\alpha\varepsilon A} = \mathcal{F}$ be a family of functions. We say $\mathcal{F}$ <u>separates points</u> if for each pair $x,y \varepsilon X$, $x \neq y$, there is an $f_\alpha \varepsilon \mathcal{F}$ with $f_\alpha(x) \neq f_\alpha(y)$. We say $\mathcal{F}$ <u>separates points from closed sets</u> if for each closed set $F \subset X$ and each $x \varepsilon X \setminus F$ there is an $f_\alpha \varepsilon \mathcal{F}$ with $f_\alpha(x) \notin \overline{f_\alpha(F)}$. The <u>evaluation function determined by $\mathcal{F}$</u> is the function $e_{\mathcal{F}} : X \to \prod_{\alpha\varepsilon A} Y_\alpha$ defined by $[e_{\mathcal{F}}(x)](\alpha) = f_\alpha(x)$.

<u>1.24 Theorem</u>. (i) If each f_α is continuous, so is $e_{\mathcal{F}}$.

(ii) If $\mathcal{F}$ separates points, $e_{\mathcal{F}}$ is $1 - 1$.

(iii) If $\mathcal{F}$ separates points from closed sets, $e_{\mathcal{F}}$ is open.

(iv) If all f_α are continuous and some f_α is a homeomorphism, $e_{\mathcal{F}}$ is an homeomorphism (into).

<u>Proof</u>. (i) $\pi_\alpha \circ e_{\mathcal{F}} = f_\alpha$ so continuity of $e_{\mathcal{F}}$ follows from 1.19.

(ii) If $f_\alpha(x) \neq f_\alpha(y)$ then $[e_{\mathcal{F}}(x)](\alpha) = f_\alpha(x) \neq f_\alpha(y) = [e_{\mathcal{F}}(y)](\alpha)$.

(iii) Let U be an open set in X and let $x \varepsilon U$. There is an α for which $f_\alpha(x) \notin \overline{f_\alpha(X \setminus U)}$ so let $V = \pi_\alpha^{-1}[Y_\alpha \setminus \overline{f_\alpha(X \setminus U)}]$. Clearly $e_{\mathcal{F}}(x) \varepsilon V$. If for some $p \varepsilon X$ for which $e_{\mathcal{F}}(p) \varepsilon V$ we had $p \notin U$ then $f_\alpha(p) \varepsilon f_\alpha(X \setminus U)$. However, $e_{\mathcal{F}}(p) \varepsilon V$ which implies that $f_\alpha(p) = \pi_\alpha(e_{\mathcal{F}}(p)) \varepsilon Y_\alpha \setminus \overline{f_\alpha(X \setminus U)}$. This contradiction says that $p \varepsilon U$ if $e_{\mathcal{F}}(p) \varepsilon V$. Thus, $e_{\mathcal{F}}(p) \varepsilon e_{\mathcal{F}}(U)$ if $e_{\mathcal{F}}(p) \varepsilon V$. We have that $V \cap e_{\mathcal{F}}(X)$ is a neighborhood of $e_{\mathcal{F}}(x)$ in $e_{\mathcal{F}}(U)$. Since $e_{\mathcal{F}}(x)$ was an arbitrary point of $e_{\mathcal{F}}(U)$ we conclude that $e_{\mathcal{F}}(U)$ is open.

(iv) If some f_α is a homeomorphism then $\mathcal{F}$ separates points and separates points from closed sets.

We do not need to concern ourselves with quotient spaces extensively but there is one result we will need.

<u>1.25 Definition</u>. Let X be a topological space and let $\mathcal{S}$ be a family of disjoint subsets of X. By $X/\mathcal{S}$ we mean the set of equivalence classes determined by the relation $x \sim y$ if and only if $x = y$ or there is an element $Y \varepsilon \mathcal{S}$ for which $x \varepsilon Y$ and $y \varepsilon Y$. The function $\phi\colon X \to X/\mathcal{S}$ is defined by $\phi(x) = [x]$, the equivalence class of x. We define a topology on $X/\mathcal{S}$ by saying a set O is open if and only if $\phi^{-1}(O)$ is open in X.

1.26 Lemma. $f\colon X/\mathcal{S} \to Y$ is continuous if and only if $f \circ \phi$ is continuous.

Proof. If f is continuous so is $f \circ \phi$. Conversely if $f \circ \phi$ is continuous and 0 is open in Y then $(f \circ \phi)^{-1}(0) = \phi^{-1}(f^{-1}(0))$ is open in X. Thus, $f^{-1}(0)$ is open.

1.27 Theorem. If $f\colon X \to Y$ is continuous, open, and **onto, then if** $\mathcal{S} = \{f^{-1}(y) | y \in Y\}$ we have $X/\mathcal{S}$ is homeomorphic to Y. **(Also,** if f is closed.)

Proof. Define $h\colon X/\mathcal{S} \to Y$ by $h([x]) = f(x)$. This is well-defined since if $[y] = [x]$ then $f(x) = f(y)$. $h \circ \phi = f$ so by 1.26 h is continuous. If $h([x]) = h([z])$ then $f(x) = f(z)$ so that $[x] = [z]$. Thus h is $1-1$.

If U is an open set in $X/\mathcal{S}$ then $h(U) = f(\phi^{-1}(U))$ and f is open. Thus $h(U)$ is open.

One final concept which we will use frequently is that of an extension of a mapping:

1.28 Definition. Let $X_o \subset X$ and $f\colon X_o \to Y$ be continuous. If there is a continuous $g\colon X \to Y$ such that $g|_{X_o} = f$, we say g is an extension of f.

1.29 Proposition. If $f\colon X_o \to Y$ has an extension $g\colon X \to Y$ where X_o is dense in X and Y is Hausdorff, then g is unique.

Proof. If $h\colon X \to Y$ is also an extension of f then $\{x \in X | g(x) = h(x)\}$ is a closed subset of X containing the dense set X_o. Hence it is all of X so that $g = h$.

The following will also prove useful.

1.30 Theorem. If $f\colon X \to Y$ is continuous and D is a dense subset of X for which $f|_D$ is a homeomorphism then if X is Hausdorff we have $f(X \setminus D) \subset Y \setminus f(D)$.

Proof. Suppose $f(x) = f(d)$ for $x \in X \setminus D$ and $d \in D$. Let V be a closed neighborhood of d which does not contain x. Then $f(V \cap D) = U \cap f(D)$ where U is open in Y. Any neighborhood 0 of x which does not intersect V does intersect D since D is dense. Because $f|_D$ is a homeomorphism it must happen that $f(0)$ is not contained in U. Since every neighborhood of x contains one which misses

V we have f is not continuous at x: no neighborhood of x is taken by f into U, a neighborhood of $f(x)$.

<u>1.31 Definition</u>. If every $f \in C(A)$ has a continuous extension to $X \supset A$, we say that A is <u>C-embedded in X</u>. Similarly, if each $f \in C^*(A)$ has a continuous extension to $X \supset A$, we say that A is <u>C*-embedded in X</u>. Note that the extension in this latter case need not be bounded: If $f(A) \subset [m,M]$ and $g: X \to \mathbb{R}$ is a continuous extension of f, then $h = (\underline{m} \vee g) \wedge \underline{M}$ is a bounded continuous extension of f.

§2. COMPACTIFICATIONS - CONSTRUCTIONS

2.1 CONVENTION. All topological spaces will be completely regular unless we specifically state otherwise. Of course, if a space is constructed, we must show that it is completely regular.

2.2 Definition. A compactification αX of the space X is a compact (Hausdorff) space αX and an embedding (homeomorphism into) $\alpha\colon X \to \alpha X$ so that $\alpha(X)$ is dense in αX. Normally, we will not distinguish between X and $\alpha(X)$. For compactifications αX, γX we say that $\alpha X \geq \gamma X$ if there is a continuous $f\colon \alpha X \to \gamma X$ for which $f \circ \alpha = \gamma$. If such an f exists which is a homeomorphism we say $\alpha X \approx \gamma X$. **($\alpha X$ is equivalent to γX.)**

2.3 Lemma. $\alpha X \approx \gamma X$ if and only if $\alpha X \geq \gamma X$ and $\gamma X \geq \alpha X$.

Proof. That $\alpha X \approx \gamma X$ implies $\alpha X \geq \gamma X$ and $\gamma X \geq \alpha X$ is easy to show: if $h\colon \alpha X \to \gamma X$ is a homeomorphism for which $h \circ \alpha = \gamma$ then $\alpha X \geq \gamma X$ and $\alpha = h^{-1} \circ \gamma$ so that $\gamma X \geq \alpha X$.

If $f\colon \alpha X \to \gamma X$, $f \circ \alpha = \gamma$, $g\colon \gamma X \to \alpha X$, and $g \circ \gamma = \alpha$ then $f \circ g \circ \gamma = f \circ \alpha = \gamma$ and since γ is a homeomorphism, it follows that $(f \circ g)|_X$ is the identity mapping. Similarly, $(g \circ f)|_X$ is the identity mapping. Thus, $f\colon \alpha X \to \gamma X$ is $1 - 1$ from a compact space onto a Hausdorff space and so is a homeomorphism (1.11, iv).

2.4 General Construction. For each $f \in C^*(X)$ let I_f be a closed interval in $\mathbb{R}$ containing $f(X)$. Any subset $F \subset C^*(X)$ which separates points and which separates points from closed sets determines an embedding $e_F\colon X \to \prod_{f \in F} I_f$ (1.24). Let $e_F X = \overline{e_F(X)}$. Then $e_F X$ is a compactification of X.

2.5 Theorem. If αX is any compactification of X then there is a subset $F \subset C^*(X)$ for which $e_F X \approx \alpha X$.

Proof. Let $F = \{f \in C^*(X) \mid \text{there is a map } f^\alpha\colon \alpha X \to \mathbb{R} \text{ with } f^\alpha \circ \alpha = f\}$.

We show first that F separates points from closed sets (and hence separates points): Let F be a closed set in X and $x \in X \setminus F$. There is a closed set $K \subset \alpha X$ such that $K \cap X = F$ and so $x \notin K$. There is then a continuous f^α: $\alpha X \to \mathbb{R}$ such that $f^\alpha(x) \notin \overline{f^\alpha(K)}$. Let $f = f^\alpha \circ \alpha$. Then $f \in C^*(X)$ and $f(x) \notin \overline{f(F)}$.

$e_F X$ is thus a compactification of X. Define h: $\alpha X \to \prod_{f \in F} I_f$ by

$$[h(p)](f) = f^\alpha(p).$$

$\pi_f \circ h = f^\alpha$ so that h is continuous (1.19). Also, $[h(\alpha(x))](f) = f^\alpha(\alpha(x)) = f(x) = e_F(x)$. This implies that $h(\alpha X)$ is a compact subset of $\prod_{f \in F} I_f$ which contains $e_F(X)$. Thus $h(\alpha X) = e_F X$. If $h(p) = h(q)$ then $[h(p)](f) = f(p) = [h(q)](f) = f(q)$ for all $f \in F$. Since F separates points, it must follow that $p = q$; i.e. h is $1 - 1$.

We have shown that h is a continuous, $1 - 1$ map from the compact space αX onto a Hausdorff space $e_F X$ and $h \circ \alpha = e_F$. We conclude that $\alpha X \approx e_F X$.

2.6 Remarks. We have shown that every family $F \subset C^*(X)$ which separates points from closed sets determines a compactification $e_F X$ of X and every compactification of X is equivalent to one of these. We can therefore consider the family of equivalence classes (under $\approx$) of the set $\{e_F X \mid F \subset C^*(X)$ and F separates points from closed sets$\}$ as the set of compactifications of X, $K(X)$, where we do not distinguish between equivalent compactifications.

2.7 Theorem (Lubben [1941]). $K(X)$ is a complete upper semi-lattice with respect to the partial order $\leq$.

Proof. Let $\{\alpha_i X\}_{i \in I}$ be a subset of $K(X)$.

We must show that this set has a least upper bound with respect to $\leq$. Define e: $X \to \prod_{i \in I} \alpha_i X$ by $[e(x)](i) = \alpha_i(x)$. Since each α_i is a homeomorphism it follows that e is also (1.24). Thus $eX = \overline{e(X)}$ is a compactification of X.

For each $i \in I$ let f_i: $eX \to \alpha_i X$ be the restriction of the projection map to eX. Then $(f_i \circ e)(x) = [e(x)](i) = \alpha_i(x)$ so that $f_i \circ e = \alpha_i$ and thus $eX \geq \alpha_i X$ for all $i \in I$.

Suppose $e_1 X \geq \alpha_i X$ for all $i \in I$, where g_i: $e_1 X \to \alpha_i X$ with $g_i \circ e_1 = \alpha_i$. Define

$$f: e_1 X \to \prod_{i \in I} \alpha_i X \text{ by } [f(p)](i) = g_i(p).$$

Then $\pi_i \circ f = g_i$ so that f is continuous and also

$f[e_1(x)](i) = g_i[e_1(x)] = \alpha_i(x) = [e(x)](i)$. We conclude that $f \circ e_1 = e$ so that $f(e_1X) = eX$ and $e_1X \geq eX$. Therefore, eX is the least upper bound of $\{\alpha_i X\}_{i\varepsilon I}$.

2.8 Corollary. $K(X)$ has a largest element, βX.

Proof. Let $\{\alpha_i X\}_{i\varepsilon I} = K(X)$.

2.9 Definition. If $f\colon X \to Y$ is continuous and $\alpha X \varepsilon K(X)$, we say that f has an extension to αX, $f^\alpha\colon \alpha X \to Y$, if $f^\alpha \circ \alpha = f$, f^α continuous. (Alternatively, $f^\alpha|_X = f$.) What we are really saying is that $f \circ \alpha^{-1}\colon X \to Y$ has an extension $f^\alpha\colon \alpha X \to Y$. If f^α exists it will be unique by 1.29 since X is dense in αX.

For $\alpha X \varepsilon K(X)$ let $C_\alpha = \{f \varepsilon C^*(X) | f$ has an extension $f^\alpha\colon \alpha X \to \mathbb{R}\}$.

2.10 Theorem. If $F \subset G \subset C^*(X)$ and F separates points from closed sets in X then $e_F X \leq e_G X$. $\alpha X \leq \gamma X$ if and only if $C_\alpha \subset C_\gamma$.

Proof. Let $h\colon \prod_{g\varepsilon G} I_g \to \prod_{f\varepsilon F} I_f$ be the projection map π_F (as in 1.20) and let $\phi = h|_{e_G X}$. It is easily verified that $\phi\colon e_G X \to e_F X$ and $g \circ e_G = e_F$. Thus, $e_F X \leq e_G X$. This shows also that $\alpha X \leq \gamma X$ if $C_\alpha \subset C_\gamma$.

Conversely, if $h\colon \gamma X \to \alpha X$ with $h \circ \gamma = \alpha$ and if $f \varepsilon C_\alpha$ with extension $f^\alpha\colon \alpha X \to \mathbb{R}$ then $f^\gamma = f^\alpha \circ h$ is an extension of f to γX.

2.11 Corollary. $\beta X \approx e_{C^*}X$.

Proof. Clearly, $\beta X \geq e_{C^*}X$ by definition.
However, $C_\beta \subset C^*(X)$ so by 2.10 $e_{C^*}X \geq \beta X$. By 2.3 $\beta X \approx e_{C^*}X$.

2.12 Construction. Let $\alpha X \varepsilon K(X)$ and suppose $\{f_i\colon X \to Y_i\}_{i\varepsilon I}$ is a family of continuous mappings, where each Y_i is compact. Since α is a homeomorphism, the evaluation map c_F determined by the family $\{\alpha\} \cup \{f_i\}_{i\varepsilon I} = F$ is an embedding of X into the compact (1.18) space $\alpha X \times \prod_{i\varepsilon I} Y_i$. $c_F X = \overline{c_F(X)}$ is a compactification of X.

2.13 Theorem. Each $f_i\colon X \to Y_i$ has a continuous extension $\overline{f}_i\colon c_F X \to Y_i$. Also, $c_F X \geq \alpha X$.

Proof. Let $\overline{f}_i$ be the restriction of the projection mapping $\pi_i\colon \alpha X \times \prod_{i\varepsilon I} Y_i \to Y_i$ to $c_F X$. $\overline{f}_i \circ c_F = f_i$, so $\overline{f}_i$ extends f_i.

Let $\overline{\alpha}$: $c_{\mathcal{F}}X \to \alpha X$ be the restriction of the projection map π_α: $\alpha X \times \prod_{i \in I} Y_i \to \alpha X$ to $c_{\mathcal{F}}X$. Then $\overline{\alpha} \circ c_{\mathcal{F}} = \alpha$ so that $c_{\mathcal{F}}X \geq \alpha X$.

<u>2.14 Corollary</u>. If f: $X \to Y$ is continuous and Y is compact, then f has an extension f^β: $\beta X \to Y$.

<u>Proof</u>. Construct $c_{\mathcal{F}}X$ as in 2.13 where $\mathcal{F} = \{\beta, f\}$. Then f has an extension $\overline{f}$: $c_{\mathcal{F}}X \to Y$ and $c_{\mathcal{F}}X \geq \beta X$. However, $\beta X \geq c_{\mathcal{F}}X$ (2.8) so that $\beta X = c_{\mathcal{F}}X$. Let $f^\beta = \overline{f}$.

<u>2.15 Remarks</u>. The first to construct a (Hausdorff) compactification for a general completely regular space was Tychonoff [1930]. His construction was essentially that which we used to obtain $e_{C^*}X$. This technique was studied extensively by Čech [1937]. It was this paper which apparently established the presently universal notation of βX for the compactification.

Using entirely different techniques, Stone [1937] constructed a compactification equivalent to βX and showed that it had the property of 2.14. Stone's construction was simplified by Gelfand-Shilov [1941] and further elaborated and generalized by Hewitt [1948].

Through some perversity of fate (perhaps misinterpretation of remarks by Hewitt [1948]) Tychonoff's name is not associated with βX, the <u>Stone-Čech</u> compactification of X, even though his paper preceeds Stone's and Cech's by seven years and is cited in both.

It may be instructive to indicate a direct proof of 2.14 which uses only the property of βX that each $f \in C^*(X)$ has an extension to βX:

<u>Alternative proof of 2.14</u>. Define f^*: $\prod_{g \in C^*(X)} I_g \to \prod_{h \in C^*(Y)} I_h$ by $[f^*(p)](h) = p(h \circ f)$.

(Note: p: $C^*(X) \to \cup I_f$ and since h: $Y \to \mathbb{R}$ we have $h \circ f \in C^*(X)$. Thus it is meaningful to write $p(h \circ f)$.)

f^* is continuous since $\pi_h \circ f^* = \pi_{h \circ f}$. Let $\overline{f} = f^*|_{\beta X}$, where $\beta X = e_{C^*}X \subset \prod_{g \in C^*(X)} I_g$. For $x \in X$ we have
$[\overline{f} \circ \beta_X(x)](h) = [\beta_X(x)](h \circ f) = [h \circ f](x)$ and
$[(\beta_Y \circ f)(x)](h) = h(f(x)) = [h \circ f](x)$ for each $h \in C^*(Y)$. Thus $\overline{f}$: $\beta X \to \beta Y$. However, βY is homeomorphic to Y, since Y is compact. Define f^β: $\beta X \to Y$ by $f^\beta = \beta_Y^{-1} \circ \overline{f}$. Then we have that $f^\beta \circ \beta_X = \beta_Y^{-1} \circ \overline{f} \circ \beta_X = f$.

2.16 Theorem. If $\alpha X \geq \gamma X$ then γX is a quotient space of αX.

Proof. The proof 2.10 shows that we can consider the map $f: \alpha X \to \gamma X$ for which $f \circ \alpha = \gamma$ to be the restriction of the projection map $\pi_{C_\gamma}: \prod_{f \in C_\alpha} I_f \to \prod_{f \in C_\gamma} I_f$ to $\alpha X \subset \prod_{f \in C_\alpha} I_f$. Since αX is compact, f is a closed mapping. It follows from 1.27 that $\alpha X/\{f^{-1}(y) | y \in \alpha X\}$ is homeomorphic to γX.

2.17 Remarks. We will see later an example which shows that $K(X)$ is not a lattice; in particular, it is generally not possible to find a greatest lower bound for a collection of compactifications. A necessary and sufficient condition for its existence is contained in the following theorem.

2.18 Theorem. Let $\{\alpha_i X\}_{i \in I} \subset K(X)$. $\text{glb}\{\alpha_i X\}_{i \in I}$ exists if and only if $\bigcap_{i \in I} C_{\alpha_i}$ separates points and separates points from closed sets.

Proof. If $\alpha X = \text{glb}\{\alpha_i X\}_{i \in I}$ then $\alpha_i X \geq \alpha X$ for all i. By 2.10 $C_\alpha \subset C_{\alpha_i}$ for all $i \in I$. It follows that $C_\alpha \subset \bigcap_{i \in I} C_{\alpha_i}$ so that this latter set separates points and separates points from closed sets, since C_α does (2.5).

Conversely, suppose $C = \bigcap_{i \in I} C_{\alpha_i}$ separates points and separates points from closed sets. Then $\alpha_i X \geq e_C X$ for all $i \in I$ since $C_{\alpha_i} \supset C$. If $\alpha_i X \geq \alpha X$ for some $\alpha X \in K(X)$ and all $i \in I$, then clearly $C_\alpha \subset C$ so that $e_C X \geq \alpha X$. This says that $e_C X = \text{glb}\{\alpha_i X\}_{i \in I}$.

2.19 Theorem. (Lubben [1941]). $K(X)$ is a complete lattice if and only if X is locally compact (i.e., each point of X has a compact neighborhood).

Proof. If X is not locally compact then for any $\alpha X \in K(X)$ it must be the case that $\alpha X \setminus X$ has more than one point. In particular X cannot be open in αX for if it were then for each $x \in X$ there is a continuous $f: \alpha X \to [0,1]$ with $f(x) = 0$ and $f(\alpha X \setminus X) = \{1\}$. Thus $f^{-1}([0,1/2])$ would be a closed (and hence, compact) neighborhood of x which is contained in X. Select two points $x_1, x_2 \in \alpha X \setminus X$. Let $C \subset C^*(X)$ be defined by $C = \{f \in C_\alpha | f^\alpha(x_1) = f^\alpha(x_2)\}$.

We claim that C separates points and separates points from closed sets (in X). For suppose that F is a closed set in X and $x \notin F$.

Then there is a closed set $G \subset \alpha X$ with $G \cap X = F$. There is a function $g\colon \alpha X \to \mathbb{R}$ with $g(x) = 0$ and $g(G \cup \{x_1, x_2\}) = 1$ (since $G \cup \{x_1, x_2\}$ is a closed set and $x \notin G \cup \{x_1, x_2\}$). Let $f = g \circ \alpha$. Then $f \in C$ and $f(x) \notin \overline{f(F)}$. Since $C \subset C_\alpha$ we have that $\alpha X \geq e_C X$. There is in C_α a function h such that $h^\alpha(x_1) \neq h^\alpha(x_2)$ (let $h = k \circ \alpha$ where $k\colon \alpha X \to \mathbb{R}$ and $k(x_1) \neq k(x_2)$). If $h \in C$ then there is an $\overline{h}\colon e_C X \to \mathbb{R}$ with $\overline{h} \circ e_C = h = h^\alpha \circ \alpha$. Let $\pi\colon \prod_{f \in C_\alpha} I_f \to \prod_{f \in C} I_f$ be the projection map. Then

$$\pi\colon \alpha X \to e_C X \text{ since } \alpha X = e_{C_\alpha} X.$$

Since for all $f \in C$ we have $f^\alpha(x_1) = f^\alpha(x_2)$ it follows that $\pi(x_1) = \pi(x_2)$. However, $\overline{h} \circ \pi = h^\alpha$ since $\overline{h} \circ \pi \circ \alpha = h \circ e_C = h = h^\alpha \circ \alpha$. (This says that $\overline{h} \circ \pi$ and h^α agree on the dense set $X \subset \alpha X$.) We have the following contradiction: $h^\alpha(x_1) \neq h^\alpha(x_2)$ but $\overline{h} \circ \pi(x_1) = \overline{h} \circ \pi(x_2)$. We conclude that $h \notin C$ and so that $\alpha X \neq e_C X$.

We have (by assuming that X is not locally compact) determined for each $\alpha X \in K(X)$ an element $e_C X \in K(X)$ with $\alpha X > e_C X$. We conclude that no glb exists for $K(X)$; i.e. $K(X)$ is not complete.

Conversely, suppose X is locally compact. Let $C = \{f \in C^*(X) \mid$ for $\varepsilon > 0$ there is a compact $K_\varepsilon \subset X$ and $|f(x)| < \varepsilon$ for all $x \in X \setminus K_\varepsilon\}$. C separates points from closed sets: Let F be closed in X and $x \in X \setminus F$. Let V be the interior of a compact neighborhood of x with $V \cap F = \phi$. Let $f\colon X \to [0,1]$ with $f(x) = 1$ and $f(X \setminus V) = \{0\}$. Clearly $f \in C$ and $f(x) \notin \overline{f(F)}$.

We claim that $C \subset C_\alpha$ for any $\alpha X \in K(X)$: For each $f \in C$, define $f^\alpha(p) = 0$ for all $p \in \alpha X \setminus X$. If $p = x$, define $f^\alpha(p) = f(x)$. We must show that f^α is continuous. Clearly f^α is continuous on X since $f^\alpha = f \circ \alpha^{-1}$ on X. For $p \in \alpha X \setminus X$ and $\varepsilon > 0$, there is a compact set $K \subset X$ such that $|f(x)| < \varepsilon$ for $x \in X \setminus K$. Let $U = \alpha X \setminus K$. Then U is a neighborhood of p and for $q \in U$ we have $|f^\alpha(q) - f^\alpha(p)| = |f^\alpha(q)| < \varepsilon$. Therefore f^α is continuous at p.

Since $C \subset C_\alpha$ for any $\alpha X \in K(X)$ it follows that $C \subset \bigcap_{i \in I} C_{\alpha_i}$ for any subset $\{\alpha_i X\}_{i \in I} \subset K(X)$. By 2.18, $\text{glb}\{\alpha_i X\}_{i \in I}$ exists.

<u>2.20 Remarks</u>. The least element of $K(X)$ (if it exists) will be denoted by ωX. We will give below an alternate construction for ωX, the

(Alexandroff) one-point compactification of X (**Alexandroff** [1924]). This compactification can be constructed for any (non-compact) space; it is Hausdorff if and only if X is locally compact and Hausdorff. This is precisely the condition we determined was necessary and sufficient for $K(X)$ to be a complete lattice. (Note that if $K(X)$ has a least element ωX then it is complete since for any family $\{\alpha_i X\}_{i\varepsilon I} \subset K(X)$ we would have $C_\omega \subset \bigcap_{i\varepsilon I} C_{\alpha_i}$ so that this latter set separates points from closed sets. Thus, by 2.18 $\{\alpha_i X\}_{i\varepsilon I}$ has a glb.)

The set $C = \{f \varepsilon C^*(X) | \text{there is an } \varepsilon > 0 \text{ and a compact } K_\varepsilon \subset X \text{ with } |f(x)| < \varepsilon \text{ if } x \varepsilon X\setminus K_\varepsilon\}$ (since it separates points from closed sets) determines a compactification $e_C X$. We showed that for any $\alpha X \varepsilon K(X)$, $C \subset C_\alpha$. Thus $\alpha X \geq e_C X$ so that $e_C X = \omega X$. Observe, however, that $C \neq C_\omega$ since this latter set consists of functions "constant at infinity": functions $f \varepsilon C^*(X)$ for which $\varepsilon > 0$, $r \varepsilon \mathbb{R}$, $K_{\varepsilon,r}$ compact in X exist such that $|f(x) - r| < \varepsilon$ for $x \varepsilon X\setminus K_{\varepsilon,r}$. ($C$ consists of functions in $C^*(X)$ which are "zero at infinity".) This characterization of C_ω is easily obtained from the alternate construction of ωX:

2.21 Construction. Let X be a locally compact space and let ∞ be a point not in X. Give $\omega X = X \cup \{\infty\}$ the topology consisting of all sets originally open in X together with any set in ωX containing ∞ whose complement was originally compact in X. As any open covering of ωX will have a set containing ∞ and thus whose complement is compact we can find a finite subcover. Therefore ωX is compact. Let ω: $X \to \omega X$ be the inclusion mapping. It is easily verified that ω is a homeomorphism and X is dense in ωX if X is not compact. Thus ωX is a compactification of X.

2.22 Proposition. $\alpha X \geq \omega X$ for all $\alpha X \varepsilon K(X)$.

Proof. Define f: $\alpha X \to \omega X$ by $f(\alpha(x)) = \omega(x)$ if $x \varepsilon X$ and by $f(p) = \infty$ for all $p \varepsilon \alpha X\setminus\alpha(X)$.

2.23 Proposition. If $\alpha X \varepsilon K(X)$ and $\alpha X\setminus X = \{p\}$ then $\alpha X \approx \omega X$.

Proof. $\alpha X \geq \omega X$ by 2.22. Define g: $\omega X \to \alpha X$ by $g(\omega(x)) = \alpha(x)$ and $g(\infty) = p$. It is easily verified that g shows $\omega X \geq \alpha X$. By 2.3, $\alpha X \approx \omega X$.

<u>2.24 Theorem</u>. Let $\{f_i: X \to Y_i\}_{i\in I}$ be a family of continuous maps, Y_i compact for each $i \in I$, and X locally compact. The compactification $c_F X$ of 2.12 is the least element of $K(X)$ to which each f_i has an extension if and only if $\alpha X = \omega X$.

<u>Proof</u>. If $\gamma X \in K(X)$ and each f_i has an extension $\overline{f}_i: \gamma X \to Y_i$, define $g: \gamma X \to \omega X \times \prod_{i\in I} Y_i$ as the evaluation map of the family $F = \{f\} \cup \{\overline{f}_i\}_{i\in I}$ where $f: \gamma X \to \omega X$ is the quotient map. Then, as before, $g(\gamma X) = c_F X$ and $g \circ \gamma = c_F$ so that $\gamma X \geq c_F X$.

If $\alpha X \neq \omega X$, let $F = \{\alpha\} \cup \{f_i\}_{i\in I}$ and $G = \{\omega\} \cup \{f_i\}_{i\in I}$. The map $h: \alpha X \times \prod_{i\in I} Y_i \to \omega X \times \prod_{i\in I} Y_i$ defined coordinate-wise by the quotient map $f: \alpha X \to \omega X$ on the first coordinate and the identity map $id_i: Y_i \to Y_i$ on the others carries $c_F X$ onto $c_G X$ and shows that $c_F X \geq c_G X$. Let $p_1 \neq p_2$ be two points of $\alpha X \setminus X$ and let $\pi_1: c_F X \to \alpha X$ be the restriction of the projection map on the first coordinate. Then $\pi_1 \circ c_F = \alpha$ so that by 1.30 there are points $q_1 \neq q_2 \in c_F X \setminus X$ with $\pi_1(q_j) = p_j$, $j = 1,2$. Since π_1 is a projection map it follows that the first coordinates of q_1 and q_2 are different. If h were a homeomorphism, then it would carry q_1 and q_2 to distinct points of $c_G X$ with distinct first coordinates (since on the other coordinates h is the identity map). By 1.30 again $h(q_1), h(q_2) \in c_G X \setminus X$ and if π is the restriction of the projection map of $\omega X \times \prod_{i\in I} Y_i$ onto ωX to $c_G X$ then $\pi(h(q_1)) \neq \pi(h(q_2))$. However, by 1.30 again $\pi(h(q_1)), \pi(h(q_2)) \in \omega X \setminus X$. This contradiction assures us that h is not a homeomorphism so that $c_F X \neq c_G X$.

<u>2.25 Example</u>. Let $f: \mathbb{R} \to [-1,1]$ be given by $f(x) = \sin x$. $\omega\mathbb{R} = S^1$, the circle, where the map $\omega: \mathbb{R} \to S^1$ is given by the inverse of the stereographic projection from the "north pole":

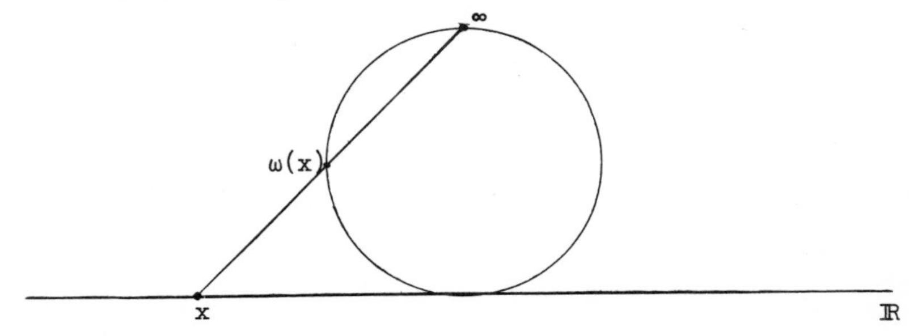

The compactification $c_F X$ where $F = \{\omega, f\}$ might be called the "closed interval compactification" of $\mathbb{R}$. It is pictured in $\omega\mathbb{R} \times [-1,1]$ as follows:

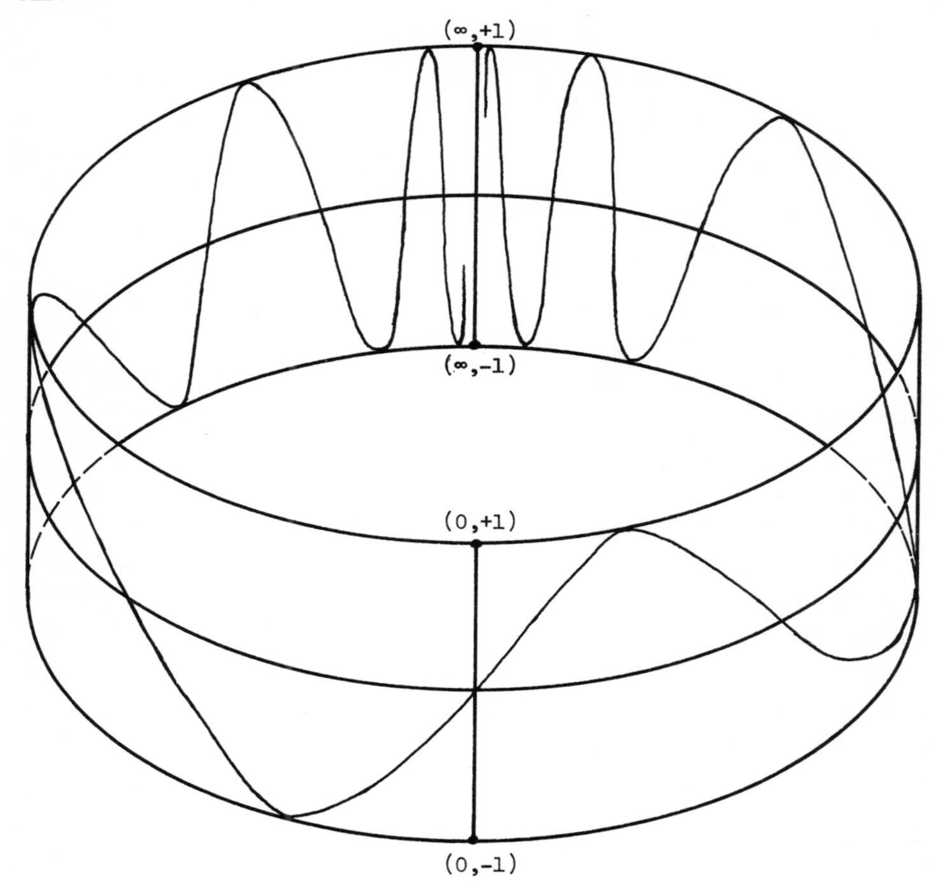

§3: ALTERNATIVE CONSTRUCTIONS OF βX

In Chapter 2 we exploited two different constructions of βX: $e_{C^*}X$ and the largest element of $K(X)$. The first of these is basically the Tychonoff-Čech definition and the second is the approach used by Engelking [1968]. We will next give the method used in Gillman and Jerison [1960]. This is an adaptation of the technique of Gelfand and **Kolmogoroff** [1939] which in turn was suggested by Stone's original construction. We will need a few preliminaries.

3.1 Definitions. A z-filter on X is a collection $\mathcal{F}$ of zero sets of X with the properties

(i) $\phi \notin \mathcal{F}$

(ii) $Z_1, Z_2 \in \mathcal{F}$ implies $Z_1 \cap Z_2 \in \mathcal{F}$

(iii) If Z is a zero set of X and $Z \supset Z_1$, $Z_1 \in \mathcal{F}$, then $Z \in \mathcal{F}$. If, in addition, the following condition is satisfied, we say $\mathcal{F}$ is a z-ultrafilter.

(iv) $\mathcal{F}$ is not properly contained in a family of zero sets which has properties (i), (ii), and (iii).

Let $\mathcal{F}$ be a z-ultrafilter on X and let M be a maximal ideal in $C(X)$.

$$Z[M] = \{Z(f) \mid f \in M\},\quad Z^{-1}[\mathcal{F}] = \{f \mid Z(f) \in \mathcal{F}\}.$$

3.2 Lemma. (a) $Z[M]$ is a z-ultrafilter on X.

(b) $Z^{-1}[\mathcal{F}]$ is a maximal ideal in $C(X)$.

Proof. (a) (i) $\phi \notin Z[M]$ since no element of M is invertible.

(ii) $Z(f) \cap Z(g) = Z(f^2 + g^2)$ so that if $f,g \in M$ so does $f^2 + g^2$.

(iii) If $Z(g) \supset Z(f)$ where $f \in M$, then $Z(g) = Z(g\cdot f)$ and $g\cdot f \in M$. Thus $Z(g) \in Z[M]$.

(iv) Suppose $Z(f) \notin Z[M]$. There must exist a $g \in M$ such that $Z(f) \cap Z(g) = \phi$; otherwise, the ideal generated by M and f is proper, contradicting the maximality of M: If $\underline{1} = g + h\cdot f$ for $g \in M$, $h \in C(X)$, then $Z(\underline{1}) = Z(g + h\cdot f) \supset Z(g) \cap Z(f) \neq \phi$, nonsense. Thus, we cannot add to $Z[M]$ any zero set which it does not already contain without violating conditions (ii) and (i) which we have already established. Hence, $Z[M]$ is a z-ultrafilter.

(b) (i) $\phi \notin \mathcal{F}$ so $\underline{1} \notin Z^{-1}[\mathcal{F}]$ Thus $Z^{-1}[\mathcal{F}]$ is a proper subset of $C(X)$.

(ii) If $f,g \in Z^{-1}[\mathcal{F}]$ then $Z(f + g) \supset Z(f) \cap Z(g)$ and by (ii) $Z(f + g) \in \mathcal{F}$. Thus $f + g \in Z^{-1}[\mathcal{F}]$.

(iii) If $f \in Z^{-1}[\mathcal{F}]$ and $g \in C(X)$ then since $Z(f\cdot g) \supset Z(f)$, we have $Z(f\cdot g) \in \mathcal{F}$. Thus $f\cdot g \in Z^{-1}[\mathcal{F}]$.

(iv) Suppose $h \notin Z^{-1}[\mathcal{F}]$. Then $Z(h) \notin \mathcal{F}$ and it follows that $Z(h) \cap Z(f) = \phi$ for some $Z(f) \in \mathcal{F}$. Thus no proper ideal in $C(X)$ can contain both f and h ($f^2 + h^2$ is an invertible element). We see that $Z^{-1}[\mathcal{F}]$ is therefore a maximal ideal in $C(X)$.

<u>3.3 Definitions</u>. An ideal $I \subset C(X)$ is a <u>z-ideal</u> if $f \in I$ whenever $Z(f) \in Z[I]$, i.e., $I = Z^{-1}Z[I]$. I is <u>prime</u> if $f\cdot g \in I$ implies $f \in I$ or $g \in I$. A z-filter $\mathcal{F}$ is <u>prime</u> if $Z_1 \cup Z_2 \in \mathcal{F}$ implies $Z_1 \in \mathcal{F}$ or $Z_2 \in \mathcal{F}$ (where, of course, Z_1 and Z_2 are zero sets on X).

<u>3.4 Lemma</u>. Let M be a maximal ideal in $C(X)$.

(i) M is a z-ideal.

(ii) M is prime.

<u>Proof</u>. (i) By 3.2 $Z^{-1}Z[M]$ is a maximal ideal which clearly contains M.

(ii) If $f\cdot g \in M$ and $f,g \notin M$ then there is an ideal M_1 properly containing M which also contains f: $M_1 = \{h\cdot f + m | h \in C, m \in M\}$. M_1 is proper since $\underline{1} \in M_1$ implies $\underline{1} = h\cdot f + m$ so that $g = \underline{1}\cdot g = h\cdot f\cdot g + m\cdot g \in M$.

<u>3.5 Lemma</u>. For a z-ideal I in $C(X)$ the following are equivalent.

(i) I is prime.

(ii) I contains a prime ideal.

<u>Proof</u>. (i) implies (ii) is trivial.

Suppose I contains a prime ideal P and $f\cdot g \in I$. Let $h = |f| - |g|$ and observe $(h \vee \underline{0})\cdot(h \wedge \underline{0}) = \underline{0}$. Thus, $h \vee \underline{0}$ or $h \wedge \underline{0} \in P$ and hence belongs to I. Say $h \vee \underline{0} \in I$. On $Z(h \vee \underline{0})$, $h \leq 0$. We see that on $Z(h \vee \underline{0})$ every zero of g is a zero of f

since $|f| \leq |g|$ there. Therefore, $Z(f) \supset Z(h \vee \underline{0}) \cap Z(f) \supset Z(h \vee \underline{0}) \cap Z(f \cdot g) \in Z[I]$. Thus, $Z(f) \in Z[I]$. Since I is a z-ideal, $f \in I$.

<u>3.6 Lemma</u>. Every prime ideal in $C(X)$ is contained in a unique maximal ideal.

<u>Proof</u>. Suppose P is prime and contained in $M_1 \cap M_2$, M_1 and M_2 maximal ideals in C. Then $M_1 \cap M_2$ is a z-ideal containing P and so $M_1 \cap M_2$ is prime. If $M_1 \neq M_2$ then there are elements $m_1 \in M_1 \setminus M_2$, $m_2 \in M_2 \setminus M_1$ so that $m_1 m_2 \in M_1 \cap M_2$ and $m_1 \notin M_1 \cap M_2$, $m_2 \notin M_1 \cap M_2$. We conclude that $M_1 = M_2$.

<u>3.7 Lemma</u>. (i) If P is a prime ideal in $C(X)$ then $Z[P]$ is a prime z-filter.

(ii) If F is a prime z-filter then $Z^{-1}[F]$ is a prime z-ideal.

<u>Proof</u>. (i) Let $P_1 = Z^{-1}Z[P]$. Then $Z[P_1] = Z[P]$ so that P_1 is a z-ideal containing the prime ideal P. By 3.5 P_1 is prime. If $Z(f) \cup Z(g) \in Z[P]$, **then** $Z(fg) = Z(f) \cup Z(g) \in Z[P_1]$ and since P_1 is a z-ideal, $fg \in P_1$. As P_1 is prime f or g is in P_1, and so $Z(f)$ or $Z(g)$ is in $Z[P_1] = Z[P]$.

(ii) Since $Z^{-1}ZZ^{-1}[F] = Z^{-1}[F]$ we conclude that $Z^{-1}[F]$ is a z-ideal. If $fg \in Z^{-1}[F]$ then $Z(fg) = Z(f) \cup Z(g) \in ZZ^{-1}[F] = F$, a prime z-filter. Thus $Z(f)$ or $Z(g) \in ZZ^{-1}[F]$ so that f or $g \in Z^{-1}[F]$.

<u>3.8 Corollary</u>. Every prime z-filter on X is contained in a unique z-ultrafilter.

<u>Proof</u>. 3.7 and 3.6.

<u>3.9 Definition</u>. A z-ultrafilter F is <u>fixed</u> if $\bigcap_{Z \in F} Z \neq \phi$.

<u>3.10 Theorem</u>. The following are equivalent:

(a) Every maximal ideal in $C^*(X)$ is fixed.

(b) Every maximal ideal in $C(X)$ is fixed.

(c) Every z-ultrafilter on X is fixed.

<u>Proof</u>. The equivalence of (b) and (c) is immediate from 3.2. (a) implies (c) since (a) says X is compact (1.14) and a z-ultrafilter is a family of closed sets in X with the finite intersection property. Finally, (b) implies (a): We show that X is compact. Then by 1.14 (a) follows.

Let $\mathcal{B}$ be any family of zero sets of X with the finite intersection property. Let $\mathcal{F}$ be the collection of zero sets of X which contain the finite intersections of members of $\mathcal{B}$. Let $\mathcal{A}$ be a z-ultrafilter containing $\mathcal{F}$. $Z^{-1}[\mathcal{A}]$ is a maximal ideal in $C(X)$ and so $\phi \neq \bigcap_{F\in\mathcal{A}} F \subset \bigcap_{F\in\mathcal{F}} F \subset \bigcap_{F\in\mathcal{B}} F$.

<u>3.11 Construction</u>. Let σX be the set of all z-ultrafilters on X. For a zero set $Z \subset X$ let $\overline{Z}$ denote all elements of σX of which Z is a member. We claim that the set $\mathcal{B} = \{\overline{Z} | Z \text{ is a zero set in } X\}$ is a base for the closed sets for a topology on σX. We must show that

(i) $\phi, \sigma X \in \mathcal{B}$

(ii) $\mathcal{B}$ is closed under finite unions.

(i) : ϕ is a zero set in X and so $\overline{\phi} \in \mathcal{B}$. However, $\overline{\phi} = \{\mathcal{F} \in \sigma X | \phi \in \mathcal{F}\} = \phi$. Therefore, $\phi \in \mathcal{B}$. Also $\overline{X} = \{\mathcal{F} \in \sigma X | X \in \mathcal{F}\} = \sigma X$ so $\sigma X \in \mathcal{B}$.

(ii) : Suppose $\overline{Z}_1, \overline{Z}_2 \in \mathcal{B}$. $Z_1 \cup Z_2 \in \mathcal{F}$ if and only if $Z_1 \in \mathcal{F}$ or $Z_2 \in \mathcal{F}$ (3.2, 3.7). Thus, the elements of σX which contain $Z_1 \cup Z_2$ are precisely those which contain Z_1 or Z_2. We see that $\overline{Z}_1 \cup \overline{Z}_2 = \overline{Z_1 \cup Z_2}$ and so $\mathcal{B}$ is closed under finite unions.

Give σX the topology having $\mathcal{B}$ as a base for the closed sets. Define σ: $X \to \sigma X$ by $\sigma(x) = \mathcal{F}_x$ where $\mathcal{F}_x = \{Z | x \in Z\}$. It is easily verified that $\mathcal{F}_x$ is a z-ultrafilter and hence belongs to σX. Now $\mathcal{F}_x \in \overline{Z} \cap \sigma(X)$ if and only if $Z \in \mathcal{F}_x$, and so, if and only if $x \in Z$. Thus, $\overline{Z} \cap \sigma(X) = \sigma(Z)$. This says that σ is a continuous and closed mapping. For $x, y \in X$ if $\sigma(x) = \sigma(y)$ then $\mathcal{F}_x = \mathcal{F}_y$ so that every zero set containing x also contains y; i.e., $x = y$. We have shown σ: $X \to \sigma X$ is a topological embedding.

We saw above that $\sigma(Z) = \overline{Z} \cap \sigma(X)$. Therefore $\text{cl}_{\sigma X}(\sigma(Z)) \subset \overline{Z}$. For any basic closed set $\overline{Z}_1$ containing $\sigma(Z)$ it follows that $\sigma(Z_1) = \overline{Z}_1 \cap \sigma(X) \supset \sigma(Z)$. Thus, $\overline{Z}_1 \supset \overline{Z}$ and so $\text{cl}_{\sigma X}(\sigma(Z)) \supset \overline{Z}$. We have proved that $\text{cl}_{\sigma X}(\sigma(Z)) = \overline{Z}$. This gives us that $\text{cl}_{\sigma X}(\sigma(X)) = \overline{X} = \sigma X$ so that $\sigma(X)$ is dense in σX.

To show that σX is a compactification of X we still need to prove that it is a compact Hausdorff space. Consider distinct points $\mathcal{F}, \mathcal{G} \in \sigma X$. There are disjoint zero sets $Z(f) \in \mathcal{F}$, $Z(g) \in \mathcal{G}$. (If every member of $\mathcal{F}$ met every member of $\mathcal{G}$ we could obtain a z-filter containing both.) Define h: $X \to [0,1]$ by

$$h(x) = |f(x)| / (|f(x)| + |g(x)|).$$

Then $h(Z(f)) = 0$ and $h(Z(g)) = 1$. Let $Z_1 = \{x \in X | h(x) \leq \frac{1}{2}\}$ and $Z_2 = \{x \in X | h(x) \geq \frac{1}{2}\}$. Then X is the union of these two zero sets, $Z(f) \cap Z_2 = \phi$, $Z(g) \cap Z_1 = \phi$. Thus, $Z_2 \notin \mathcal{F}$ and $Z_1 \notin \mathcal{G}$. Since $cl_{\sigma X}(\sigma(Z_i)) = \overline{Z}_i$, $i = 1,2$, it follows that $\mathcal{F} \notin cl_{\sigma X}(\sigma(Z_2))$ and $\mathcal{G} \notin cl_{\sigma X}(\sigma(Z_1))$. The open sets $O_i = \sigma X \setminus cl_{\sigma X}(\sigma(Z_i))$ are disjoint ($Z_1 \cup Z_2 = X$ so that $\overline{Z}_1 \cup \overline{Z}_2 = \sigma X$; **therefore** $O_1 \cap O_2 = \phi$) and $\mathcal{F} \in O_2$, $\mathcal{G} \in O_1$.

To show σX is compact, consider any family $\{\overline{Z}\}_{Z \in \mathcal{B}}$ of basic closed sets with the finite intersection property. For any finite set $Z_1, \ldots, Z_n \in \mathcal{B}$ if $\mathcal{F}$ is an element of $\overline{Z}_1 \cap \ldots \cap \overline{Z}_n$ then $Z_1, \ldots, Z_n \in \mathcal{F}$. Thus $Z_1 \cap \ldots \cap Z_n \neq \phi$. We see that $\mathcal{B}$ has the finite intersection property. Let $\mathcal{A}$ be the family of zero sets which contain all finite intersections of members of $\mathcal{B}$. It is easily verified that $\mathcal{A}$ is a z-ultrafilter and $\mathcal{A} \in \bigcap_{Z \in \mathcal{B}} \overline{Z}$.

<u>3.12 Theorem</u>. $\sigma X \approx \beta X$.

<u>Proof</u>. We show that $\sigma X \geq \beta X$. For $\mathcal{F} \in \sigma X$ let $\mathcal{G}_{\mathcal{F}} = \{Z_\beta \subset \beta X | \text{there exist } Z \in \mathcal{F} \text{ with } Z_\beta \supset Z\}$. It is easily verified that $\mathcal{G}_{\mathcal{F}}$ is a z-filter on βX. If $Z_1 \cup Z_2 \in \mathcal{G}_{\mathcal{F}}$ then there is a $Z \in \mathcal{F}$ with $Z \subset Z_1 \cup Z_2$. $Z_1 \cap X$ and $Z_2 \cap X$ are zero sets in X whose union contains Z. Since $\mathcal{F}$ is prime, one belongs to $\mathcal{F}$, say $Z_1 \cap X$. Since Z_1 is a zero set in βX containing $Z_1 \cap X$ we conclude that $Z_1 \in \mathcal{G}_{\mathcal{F}}$.

We have shown that $\mathcal{G}_{\mathcal{F}}$ is prime. By 3.8 there is a unique z-ultrafilter $\mathcal{A}_{\mathcal{F}}$ containing $\mathcal{G}_{\mathcal{F}}$. By 3.10 $\mathcal{A}_{\mathcal{F}}$ is fixed. Thus $\bigcap_{F \in \mathcal{A}_{\mathcal{F}}} F \neq \phi$. We claim there is a unique point in this intersection. Suppose $p_1, p_2 \in F$ for all $F \in \mathcal{A}_{\mathcal{F}}$. If $p_1 \neq p_2$ there is a zero set $Z \subset \beta X$ containing p_1 and not containing p_2. Clearly $Z \notin \mathcal{A}_{\mathcal{F}}$ but $Z \cap F \neq \phi$ for all $F \in \mathcal{A}_{\mathcal{F}}$ since $p_1 \in F$. This contradicts the maximality of $\mathcal{A}_{\mathcal{F}}$. Define $f: \sigma X \to \beta X$ by $f(\mathcal{F}) = \bigcap_{F \in \mathcal{A}_{\mathcal{F}}} F$. For $x \in X$ we have $\sigma(x)$ is the family of all zero sets in X containing x so that $f(\sigma(x)) = \bigcap_{F \in \mathcal{A}_{\sigma(x)}} F \subset \bigcap_{F \in \mathcal{G}_{\sigma(x)}} F$. We claim that this latter intersection contains only $\beta(x)$. This is easily seen from the fact that $\mathcal{G}_{\sigma(x)} = \{Z_\beta \subset \beta X | \beta(x) \in Z_\beta\}$. To prove this, observe that by definition $\mathcal{G}_{\sigma(x)}$ is contained in this second set. If we have a $Z_\beta(t)$ in the second set then $Z(t \circ \beta) \in \sigma(x)$ and $Z_\beta(t) \supset \beta(Z(t \circ \beta))$. Thus, $Z_\beta(t) \in \mathcal{G}_{\sigma(x)}$.

We have now shown that $f \circ \sigma = \beta$. We need to show that f is continuous. For any $\mathcal{F} \in \sigma X$ let U be a zero set neighborhood of $f(\mathcal{F})$ in βX and let U' be a zero set in βX for which $f(\mathcal{F}) \in \beta X \setminus U' \subset U$. Let $Z = \beta^{-1}(U)$, $Z' = \beta^{-1}(U')$. Then Z, Z' are zero sets in X for which $Z \cup Z' = X$ (since $U \cup U' = \beta X$). We claim $O = \sigma X \setminus \overline{Z'}$ is a neighborhood of $\mathcal{F}$ in X for which $f(O) \subset U$. Suppose $\mathcal{F} \in \overline{Z'}$. Then $Z' \in \mathcal{F}$ and we conclude that $U' \in \mathcal{G}_{\mathcal{F}}$ $(U' \supset \beta(Z'))$. This is impossible since $f(\mathcal{F}) \notin U'$. We have shown that O is a neighborhood of $\mathcal{F}$. Suppose $\mathcal{F}' \in O$. Then $Z' \notin \mathcal{F}'$. However $Z \cup Z' = X \in \mathcal{F}'$ so $Z \in \mathcal{F}'$. U is a zero set in βX containing $\beta(Z)$. Therefore $U \in \mathcal{G}_{\mathcal{F}'}$ and so $f(\mathcal{F}') \in U$.

3.13 Remarks. Comfort [1968] showed that one could obtain βX without using the axiom of choice provided the definition of compactness was changed: A space X is called compact* if every maximal ideal in $C^*(X)$ is fixed. It is shown in Comfort [1968] and in Salbany [1974] that $e_{C^*}X$ is compact* without using the axiom of choice.

The purpose of this last construction (Chandler [1972]) is an alternate approach along these lines. We will give a different means of constructing βX and show that it is compact* without using the axiom of choice. We need some preliminary results from the above cited paper of Comfort to do this. It should be noted that what follows (through 3.19) is an alternate approach to that of Gillman and Jerison [1960], Chapter V.

3.14 Definition. Let M be a maximal ideal in $C^*(X)$. For arbitrary $f + M$, $g + M$ in $C^*(X)/M$ define $f + M < g + M$ provided there exist $\varepsilon > 0$, $\delta > 0$, $h \in M$ such that $f + \underline{\varepsilon} < g$ on $\{x \in X \mid -\delta \leq h(x) \leq \delta\} = Z_\delta(h)$.

3.15 Proposition. (i) $<$ is well-defined, antisymmetric, and transitive.

(ii) $f + M > \underline{0} + M$, $g + M > \underline{0} + M$ implies $f + g + M > \underline{0} + M$ and $fg + M > \underline{0} + M$.

(iii) $f + M > g + M$ if and only if $f - g + M > \underline{0} + M$.

(iv) $C^*(X)/M$ is a partially ordered field with respect to $<$.

(v) The map ϕ: $\mathbb{R} \to C^*(X)/M$ defined by $\phi(r) = \underline{r} + M$ is an order isomorphism of $\mathbb{R}$ onto a linearly ordered subfield of $C^*(X)/M$.

Proof. (i) If $f + p = f_1$, $g + q = g_1$, and $f + \underline{\varepsilon} < g$ on $Z_\delta(h)$ where $p,q,h \in M$, then let $k = h^2 + p^2 + q^2$, $\mu = \varepsilon/3$, $\nu = \min\{\delta^2, (\varepsilon/3)^2\}$. Then on $Z_\nu(k)$ we have $|h| < \delta$, $|p| < \varepsilon/3$, $|q| < \varepsilon/3$ so that $g_1 - f_1 = g - f + q - p > \underline{\varepsilon} - \underline{\varepsilon/3} - \underline{\varepsilon/3} = \underline{\varepsilon/3}$. Thus $f_1 + \underline{\varepsilon/3} < g_1$ and so $f_1 + M < g_1 + M$. We have shown that $<$ is well-defined.

$f + \underline{\varepsilon} < g$ on $Z_\delta(h)$ and $g + \underline{\mu} < f$ on $Z_\nu(k)$ gives an immediate contradiction since $Z_\delta(h) \cap Z_\nu(k) \neq \emptyset$. ($h, k \in M$ implies $h^2 + k^2 \in M$ and so $Z_\eta(h^2 + k^2) \neq \emptyset$ for any $\eta > 0$. Let $\eta = \min\{\delta^2, \nu^2\}$. For this choice of η, $Z_\eta(h^2 + k^2) \subset Z_\delta(h) \cap Z_\nu(k)$.) This shows that $<$ is antisymmetric.

If $f + M < g + M$ and $g + M < h + M$ then there exist $\varepsilon > 0$, $\delta > 0$, $\mu > 0$, $\nu > 0$, $k, \ell \in M$ for which $f + \underline{\varepsilon} < g$ on $Z_\delta(k)$ and $g + \underline{\mu} < h$ on $Z_\nu(\ell)$. Then $f + \underline{\rho} < h$ on $Z_\sigma(p)$ where $\rho = \varepsilon + \mu$, $\sigma = \min\{\delta^2, \nu^2\}$, $p = h^2 + \ell^2$. Thus, $f + M < h + M$ and we see that $<$ is transitive.

(ii) $f + M, g + M > \underline{0} + M$ implies that $\varepsilon > 0$, $\delta > 0$, $\mu > 0$, $\nu > 0$, $h \in M$, $k \in M$ exist for which $\underline{0} + \underline{\varepsilon} < f$ on $Z_\delta(h)$ and $\underline{0} + \underline{\mu} < g$ on $Z_\nu(\mathbf{k})$. Then $\underline{0} + \underline{\rho} < f + g$ and $\underline{0} + \underline{\pi} < fg$ on $Z_\sigma(p)$ where $\rho = \varepsilon + \mu$, $\pi = \varepsilon\mu$, $\sigma = \min\{\delta^2, \nu^2\}$, $p = h^2 + k^2$. Thus, $f + g + M > \underline{0} + M$, and $fg + M > \underline{0} + M$.

(iii) $f + M > g + M$ if and only if $\varepsilon > 0$, $\delta > 0$, $h \in M$ exist for which $g + \underline{\varepsilon} < f$ on $Z_\delta(h)$ if and only if $\underline{0} + \underline{\varepsilon} < f - g$ on $Z_\delta(h)$ if and only if $f - g + M > \underline{0} + M$.

(iv) To establish this, in view of (ii), all we must show is that $f + M \geq \underline{0} + M$ and $-f + M \geq \underline{0} + M$ implies that $f \in M$. In other words, we need to establish that $f + M > \underline{0} + M$ and $-f + M > \underline{0} + M$ are not both possible. If this were the case, we would have by (iii) that $f + (-f) + M > \underline{0} + M$. Thus, $\underline{0} + M > \underline{0} + M$, a contradiction.

(v)
$$\begin{aligned} \phi(r + s) &= \underline{r + s} + M = \underline{r} + \underline{s} + M = \underline{r} + M + \underline{s} + M \\ &= \phi(r) + \phi(s). \\ \phi(rs) &= \underline{rs} + M = \underline{r}\,\underline{s} + M = (\underline{r} + M)(\underline{s} + M) \\ &= \phi(r)\,\phi(s). \end{aligned}$$

Thus, ϕ is a ring homomorphism of $\mathbb{R}$ into C^*/M. To show it is an order isomorphism, we must show that if $r > 0$ then $\underline{r} + M > \underline{0} + M$. This is trivial since $\underline{0} + \underline{r/2} < \underline{r}$ on $X = Z_1(\underline{0})$.

<u>3.16 Lemma</u>. $(f \vee \underline{0}) + M > \underline{0} + M$ and $(-f \vee \underline{0}) + M > \underline{0} + M$ cannot both hold.

<u>Proof</u>. Let $f^+ = f \vee \underline{0}$, $f^- = -f \vee \underline{0}$. If $\varepsilon > 0$, $\delta > 0$, $\mu > 0$, $\nu > 0$, $h, k \in M$ exist with $\underline{0} + \underline{\varepsilon} < f^+$ on $Z_\delta(h)$ and $\underline{0} + \underline{\mu} < f^-$ on $Z_\nu(k)$ then $Z_\delta(h) \cap Z_\nu(k) \neq \emptyset$ (otherwise M would not be a proper ideal) so that at any point x in this intersection we would have $f^+(x) > \varepsilon$ and

$f^-(x) > \mu$ which obviously cannot happen.

<u>3.17 Lemma</u>. For any $f \in C^*(X)$ either $f \in M$ or $f^2 + M > \underline{0} + M$.

<u>Proof</u>. If $f \notin M$ then neither does f^2. Since M is maximal there are $g \in C^*$, $h \in M$ such that $\underline{1} = gf^2 + h$. If $f^2 + M \not> \underline{0} + M$ then for each positive integer n there is an $x \in Z_{1/n}(h)$ for which $f^2(x) \leq \underline{0}(x) + \underline{1/n}(x)$, i.e., $f^2(x) \leq 1/n$. For this x we must have $1 = \underline{1}(x) = g(x) \cdot f^2(x) + h(x)$. Thus, $1 \leq g(x) \cdot 1/n + 1/n$ and so $g(x) \geq n - 1$. This says that g is unbounded on X, a contradiction. We conclude that either $f \in M$ or $f^2 + M > \underline{0} + M$.

<u>3.18 Corollary</u>. If $f \geq \underline{0}$ then either $f \in M$ or $f + M > \underline{0} + M$.

<u>Proof</u>. We may construct a square root $f^{1/2}$ for f. Since M is prime then $f \in M$ if and only if $f^{1/2} \in M$. By 3.17 either $f^{1/2} \in M$ or $f + M > \underline{0} + M$.

<u>3.19 Theorem</u>. C^*/M is linearly ordered by $<$ and is order isomorphic to $\mathbb{R}$.

<u>Proof</u>. In view of 3.15 (iii) to show $<$ is a linear order on C^*/M we must establish for each $f \in C^*$ that one of the following hold: $f + M < \underline{0} + M$, $f + M > \underline{0} + M$, $f \in M$. As before, let $f^+ = f \vee \underline{0}$, $f^- = -f \vee \underline{0}$. Then $f = f^+ - f^-$ so that $f \in M$ if $f^+ \in M$ and $f^- \in M$. If $f^+ \notin M$ then by 3.18 $f^+ + M > \underline{0} + M$ and by 3.16 $f^- \in M$. Thus, $f + M = f^+ - f^- + M = f^+ + M > \underline{0} + M$. A similar argument in case $f^- + M > \underline{0} + M$ would establish that $f + M < \underline{0} + M$.

To show that C^*/M is order isomorphic to $\mathbb{R}$ (since we have already shown that it has a subfield order isomorphic to $\mathbb{R}$: 3.15 (v)) all we need verify is that C^*/M is Archimedean: For any $f \in C^*$ we have an n such that $|f| < \underline{n}$. Thus on $X = Z_1(\underline{0})$ we have $f + \underline{\varepsilon} < \underline{n + 1}$ for any $\varepsilon \leq 1$. Thus $f + M < \underline{n + 1} + M$.

<u>3.20 Definition</u>. A <u>C^*-extension</u> of X is a pair (Y,r) where $r\colon X \to Y$ is an embedding, $r(X)$ is dense in Y, and for each $f \in C^*(X)$ there is an $\bar{f} \in C^*(Y)$ such that $\bar{f} \circ r = f$. Two C^*-extensions of X, (Y_1, r_1) and (Y_2, r_2) are <u>equivalent</u> (written $(Y_1, r_1) \approx (Y_2, r_2)$) if there is a homeomorphism $h\colon Y_1 \to Y_2$ for which $h \circ r_1 = r_2$.

<u>3.21 Lemma</u>. If D is a dense subset of X then $|X| \leq 2^{2^{|D|}}$.

Proof. Define a function $\tau: X \to P(P(D))$ by $\tau(x) = \{U \cap D | U$ is a neighborhood of x in $X\}$.

We show τ is $1-1$: For $x \neq y$ choose disjoint neighborhoods U of x and V of y. If $U \cap D$ were a member of $\tau(y)$ we would have $(U \cap V) \cap D \neq \phi$. (If $U \cap D = O \cap D$ for O a neighborhood of y, then $(U \cap V) \cap D = U \cap D \cap V = O \cap D \cap V = (O \cap V) \cap D \neq \phi$.) Since $U \cap D \in \tau(x)$ and $U \cap D \notin \tau(y)$ we have $\tau(x) \neq \tau(y)$. Since τ is $1-1$ it follows that $|X| \leq |P(P(D))| = 2^{2^{|D|}}$.

3.22 Theorem. There exists a set E (obtained without the axiom of choice) of C^*-extensions of X with the property that any C^*-extension of X is equivalent to a member of E.

Proof. If (Y,r) is a C^*-extension of X there is a function $\phi: Y \to P(P(X)) = P^2(X)$ which is $1-1$. (3.21). Define a topology on $P^2(X)$ by declaring any subset of $P^2(X) \setminus \phi(Y)$ to be open and a subset $U \subset \phi(Y)$ is open if and only if $\phi^{-1}(U)$ is open in Y. Then ϕ is an embedding and we can consider Y to be a subspace of $P^2(X)$ with this topology. Let T be the set of distinct completely regular topologies on $P^2(X)$; let $E_Z = \{Z\} \times C(X,Z)$, where Z is a subspace of $(P^2(X), T)$, $T \in T$; let E_1 be the disjoint union of the E_Z over all subspaces of $P^2(X)$ with topologies from T; and finally, let E be the subset of E_1 consisting of those pairs (Z,s) which are C^*-extensions of X.

3.23 Construction. Let Λ be the disjoint union of all Y appearing as first element in a member of E. Define an equivalence relation $\sim$ on Λ by saying $y_\alpha \sim y_{\alpha'}$ provided $f_\alpha(y_\alpha) = f_{\alpha'}(y_{\alpha'})$ for all $f \in C^*(X)$ where $y_\alpha \in Y_\alpha$, $y_{\alpha'} \in Y_{\alpha'}$, (Y_α, r_α), $(Y_{\alpha'}, r_{\alpha'}) \in E$ and f_α, $f_{\alpha'}$ are the "extensions" of f to Y_α, $Y_{\alpha'}$: $f_\alpha \circ r_\alpha = f = f_{\alpha'} \circ r_{\alpha'}$. Let χX be the set of equivalence classes of Λ determined by $\sim$. For each $f \in C^*(X)$ there is a function $f_\chi: \chi X \to \mathbb{R}$ defined by $f_\chi([y_\alpha]) = f_\alpha(y_\alpha)$. It is clear that the equivalence relation on Λ makes f_χ well-defined. Give χX the weakest topology for which all the f_χ, $f \in C^*(X)$, are continuous.

3.24 Lemma. χX is completely regular.

Proof. In view of 1.6, all we need to show is that χX is Hausdorff. This is immediate since if $[y_\alpha] \neq [z_\alpha]$ there is an $f \in C^*(X)$ such that for some index, $f_\alpha(y_\alpha) \neq f_\alpha(z_\alpha)$. Then $f_\chi([y_\alpha]) \neq f_\chi([z_\alpha])$.

<u>3.25 Definition</u>. Let $\chi\colon X \to \chi X$ be defined by $\chi(x) = [r_\alpha(x)]$ where (Y_α, r_α) is some element of E. Also, define $\eta_\alpha\colon Y_\alpha \to \chi X$ by $\eta_\alpha(y_\alpha) = [y_\alpha]$.

<u>3.26 Proposition</u>. (i) $\chi\colon X \to \chi X$ is an embedding.
(ii) $\eta_\alpha\colon Y_\alpha \to \chi X$ is continuous and $f_\chi \circ \eta_\alpha = f_\alpha$ for each $f \in C^*(X)$.

<u>Proof</u>. (i) $f_\chi \circ \chi = f$ for each $f \in C^*(X)$ so if O is a basic open set in χX then $O = f_\chi^{-1}(U)$ for some $U \subset \mathbb{R}$ (open) and some $f \in C^*(X)$. Thus

$$\chi^{-1}(O) = \chi^{-1}f_\chi^{-1}(U) = (f_\chi \circ \chi)^{-1}(U) = f^{-1}(U),$$

which is open in X. Therefore, χ is continuous.

If $x \neq x_1$ in X there is some $f \in C^*(X)$ for which $f(x) \neq f(x_1)$. Then

$$f_\alpha(r_\alpha(x)) = f(x) \neq f(x_1) = f_\alpha(r_\alpha(x_1)) \quad \text{so} \quad r_\alpha(x) \not\sim r_\alpha(x_1)$$

and it follows that

$$\chi(x) = [r_\alpha(x)] \neq [r_\alpha(x_1)] = \chi(x_1).$$

We have shown χ is $1 - 1$.

Let U be an open set in X and suppose $x \in U$. There is an element $f \in C^*(X)$ for which $f(x) = 0$ and $f(X \setminus U) = \{1\}$. Then $f_\chi^{-1}([0, 1/2]) \cap \chi(X)$ is an open set containing $\chi(x)$ and contained in $\chi(U)$. Thus, χ is an open mapping.

(ii) $f_\chi \circ \eta_\alpha = f_\alpha$ follows from the definitions of the three functions. The continuity of η_α is proved in the same way as we proved the continuity of χ above.

<u>3.27 Theorem</u>. χX is compact*.

<u>Proof</u>. Suppose M is a free maximal ideal in $C^*(\chi X)$. Let $Y = \chi X \cup \{p\}$ where p is some point not in χX. For each $f \in C^*(X)$ define $\tilde{f}\colon Y \to \mathbb{R}$ by

$$\tilde{f}(y) \begin{cases} = f_\chi(y) & \text{if } y \in \chi X \\ = r_f & \text{if } y = p \end{cases}$$

where r_f is the real number carried by the isomorphism $\phi\colon \mathbb{R} \to C^*(X)/M$ to f (3.19). Give Y the weakest topology making all $\tilde{f}$, $f \in C^*(X)$, continuous. Define $r\colon X \to Y$ by $r(x) = \chi(x)$. We claim (Y,r) is a C^*-extension of X: Y is completely regular since it is Hausdorff (1.6), (Y is Hausdorff since the family $\{\tilde{f} \mid f \in C^*(X)\}$ separates points: the only points which would give any problem would be

p and some $x \in \chi X$. If $f_\chi(x) = r_f$ for all $f \in C^*(X)$ then M would be fixed since each $r_g = 0$ for $g \in M$ and each g is an f_χ for $f \in C^*(X)$; namely $g = (g \circ \chi)_\chi$.)

$r(X)$ is dense in Y since $r = i \circ \chi$ (where $i: \chi X \to Y$ is the inclusion mapping) and $i(\chi X)$ is dense in Y: $f - \underline{r}_f \in M$ implies for any neighborhood U of r_f in $\mathbb{R}$ there is a point $y \in \chi X$ for which $\tilde{f}(y) \in U$ (otherwise $f - \underline{r}_f$ is invertible). Since $\tilde{f}^{-1}(U)$ is a basic neighborhood of p in Y we have every neighborhood of p in Y contains an element of χX. The inclusion i is clearly an embedding so we have shown that $r(X)$ is dense in Y and $r = i \circ \chi$ is an embedding. Thus (Y,r) is a C^*-extension of X.

By 3.22 there is a member (Y_1, r_1) of E and a homeomorphism $h: Y \to Y_1$ with $h \circ r = r_1$. For each $f_\chi \in M$ we have $\tilde{f}(p) = 0$ where $f = f_\chi \circ \chi$ since $r_f = 0$ for those f for which $f_\chi \in M$. The commutativity of the diagram

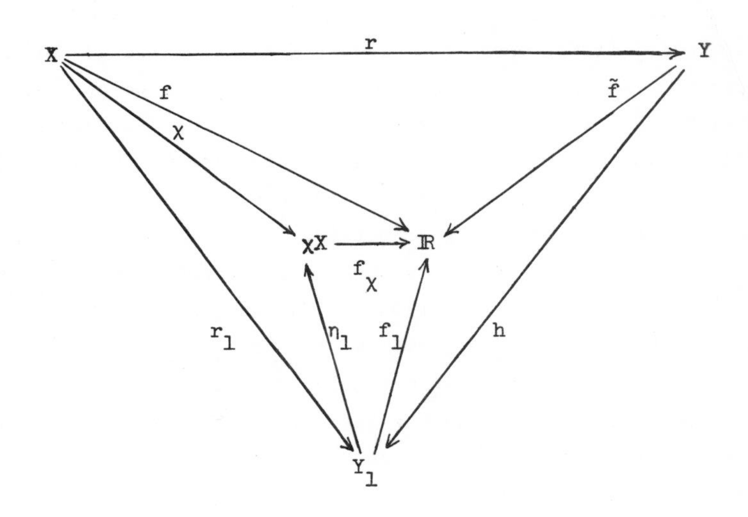

then implies that

$$f_\chi(\eta_1[h(p)]) = f_1[h(p)] = \tilde{f}(p) = 0.$$

We see then that M is fixed, a contradiction.

This says that no maximal ideals of $C^*(\chi X)$ are free. Thus X is compact*.

<u>3.28 Remarks</u>. With the axiom of choice compact* implies compact. This shows that $\chi X \approx e_{C^*}X = \beta X$ since each $f \in C^*(X)$ has an extension

$f_\chi: \chi X \to \mathbb{R}$. The proof that χX is compact* depends heavily on the fact that $C^*(X)/M \cong \mathbb{R}$ (3.19). This isomorphism does not hold if C^* is replaced by C: If M is a maximal ideal in $C(X)$ then C/M may be non-Archimedean. If $C/M \cong \mathbb{R}$ we say M is <u>real</u>. When every real maximal ideal in $C(X)$ is fixed, we say X is <u>realcompact</u>. Many of the results on compactifications have analogues for realcompactifications and, in particular, there is a realcompactification υX to which each $f \in C(X)$ has an extension. It was first constructed by Hewitt [1948]. His method is an exact analogue of the construction of $e_{C^*}X$ using C in place of C^* and $\mathbb{R}$ in place of I_f.

υX can also be constructed in exactly the same manner as χX was, using C-extensions in place of C^*-extensions. The fact that C/M is not always isomorphic to $\mathbb{R}$ gives no difficulty since to establish that the space constructed is realcompact all we need is to assume that some free maximal ideal M is <u>real</u> and obtain a contradiction.

We can also consider υX to be the largest subspace of βX in which X is C-embedded (i.e., each $f \in C(X)$ has an extension to $\upsilon X \subset \beta X$ and no subspace of βX properly containing υX has this property.) Note the analogy with βX: it is the largest subspace of βX in which X is C^*-embedded. It was these considerations which led to the above construction of χX. Of course, this had to be done without using βX; after all, that is what we are trying to obtain.

Until recently, the definitive treatise on realcompactness has been the book by Gillman and Jerison [1960]. In fact, one of the major purposes of this book is to show the interplay of topology on X with algebra on $C(X)$ and this seems most nicely realized by the realcompact spaces. There have appeared in the last year or so several books containing a significant amount of material concerning realcompactness. For example, see Weir [1975] and Comfort and Negrepontis [1975].

§4: EXAMPLES

4.1 Basic Construction: Ordinal Spaces.

We shall assume the basic facts about ordinals: They are the order types of well-ordered sets (Sierpinski [1965]) or equivalently, they are the members of the proper class On which is well-ordered by the membership relation ε (Takeuti and Zaring [1971]). We will denote ordinal numbers by lower-case Greek letters. For $\alpha \in On$, $W(\alpha) = \{\sigma \in On | \sigma < \alpha\}$. ω will denote the smallest member of On with infinitely many predecessors: $W(\omega)$ is infinite and for all $\alpha < \omega$, $W(\alpha)$ is finite. ω_1 will denote the smallest member of On with uncountably many predecessors. Following Gillman and Jerison [1960] we will denote $W(\omega)$ by N (since $W(\omega)$ is order isomorphic to the positive integers), $W(\omega_1)$ by W, and $W(\omega_1 + 1)$ by W^*. A limit ordinal λ is a member of On for which $W(\lambda)$ has no largest element. For any member $\alpha \in On$ there is a unique limit ordinal λ and a unique finite ordinal n for which $\alpha = \lambda + n$. (0 is the least member of On and is the order type of the empty set.)

We define a topology on $W(\alpha)$ by saying the basic neighborhoods of a point $\tau \in W(\alpha)$ are the intervals
$[\sigma + 1, \tau] = (\sigma, \tau + 1) = \{x \in W(\alpha) | \sigma < x < \tau + 1\}$.

4.2 Proposition. For each $\alpha \in On$, $W(\alpha)$ is normal.

Proof. For $\sigma < \tau$, $\sigma, \tau \in W(\alpha)$, $[\pi, \sigma]$ and $(\sigma, \tau + 1)$ are disjoint open sets containing σ and τ respectively, where π is any element of $W(\alpha)$ smaller than σ. (If $\sigma = 0$, then choose $\pi = \sigma$.) Thus, $W(\alpha)$ is Hausdorff.

Suppose H and K are disjoint closed sets in $W(\alpha)$. For each $\sigma \in H$, $\tau \in K$ let U_σ, V_τ be basic neighborhoods of the form $[\pi, \sigma]$, $[\rho, \tau]$ which do not meet K, H, respectively. Let $U = \bigcup_{\sigma \in H} U_\sigma$, $V = \bigcup_{\tau \in K} V_\tau$. Then $H \subset U$, $K \subset V$ and if $U \cap V \neq \emptyset$ then $[\pi, \sigma] \cap [\rho, \tau] \neq \emptyset$ for some $\sigma \in H$, $\tau \in K$. If $\sigma < \tau$ then $\sigma \in [\rho, \tau]$ and if $\tau < \sigma$ then $\tau \in [\pi, \sigma]$, in either case, a contradiction. Thus $U \cap V = \emptyset$.

4.3 Proposition. $W(\alpha)$ is compact if and only if α is not a limit ordinal.

Proof. If α is a limit ordinal, then $\{[0,\sigma)|\sigma \in W(\alpha)\}$ is an open cover of $W(\alpha)$, clearly without a finite subcover.

If $\alpha = \sigma + 1$ (i.e. σ is the immediate predecessor of α) and $\{U_i\}_{i \in I}$ is an open cover of $W(\alpha)$ then some U_o contains σ. Let σ_1 be the smallest element of the set $\{x \in W(\alpha)|(x,\sigma] \subset U_o\}$. Then $\sigma_1 \notin U_o$ so there must be U_1 containing σ_1. Let σ_2 be the smallest element of the set $\{x \in W(\alpha)|(x,\sigma_1] \subset U_1\}$. Then $\sigma_2 < \sigma_1$ and $\sigma_2 \notin U_1$. The only thing which would prevent us from constructing an infinite decreasing sequence $\sigma_1 > \sigma_2 > \sigma_3 > \ldots$ in $W(\alpha)$ (which is clearly not possible: it has no smallest element) is for some index n, the set $\{x \in W(\alpha)|(x, \sigma_n] \subset U_n\}$ has 0 as its smallest element. Thus, $W(\alpha) \subset U_o \cup U_1 \cup \ldots \cup U_n \cup U_{n+1}$, where U_{n+1} is some element of the cover containing 0.

We have shown that every open cover of $W(\alpha)$ has a finite subcover; i.e., $W(\alpha)$ is compact.

4.4 Corollary. W is not compact. W^* is compact.

4.5 Definitions, Remarks, Etc. For a set A, $|A|$ (**read** "the cardinal number of A") is the least ordinal which is equivalent to A. The proper class of infinite ordinals which are cardinal numbers will be denoted by N'. There is an order preserving isomorphism of On onto N' (Takeuti and Zaring [1971], p. 81) and the image of α under this isomorphism will be denoted $\aleph_\alpha$. The ordinal number represented by $\aleph_\alpha$ will be denoted by ω_α. (Note: Tradition has dictated a distinction between cardinal numbers, which include the alephs, and ordinal numbers which we are going to preserve here.) Thus, $\omega_o = \aleph_0 = \omega$, $\omega_1 = \aleph_1$, etc.

4.6 Lemma. If A is a subset of $W(\omega_\alpha)$ and $|A| < \aleph_\alpha$ then there is in $W(\omega_\alpha)$ a least upper bound of A.

Proof. The set $W(\omega_\alpha) \setminus \bigcup_{\sigma \in A} W(\sigma)$ cannot be empty for if it were we would have $W(\omega_\alpha)$ as the union of fewer than $\aleph_\alpha$ sets, each with fewer than $\aleph_\alpha$ members. lub A is the smallest element of $W(\omega_\alpha) \setminus \bigcup_{\sigma \in A} W(\sigma)$.

4.7 Corollary. A countable set in W has a lub.

4.8 Lemma. Of any two disjoint closed sets in $W(\omega_\alpha)$, $\alpha \geq 1$, at least one is bounded.

Proof. If H and K are disjoint, closed, unbounded sets in $W(\omega_\alpha)$, then select a sequence $\sigma_1 < \sigma_2 < \dots$ where $\sigma_{2n} \in H$, $\sigma_{2n-1} \in K$, $n = 1,2,3,\dots$. If $\sigma = \text{lub}\,\{\sigma_k\}$ then σ is a limit point of both H and K.

4.9 Definition. A space is countably compact if every countable open covering has a finite subcovering.

4.10 Theorem. $W(\omega_\alpha)$ is countably compact if $\alpha \geq 1$.

Proof. Let $\{U_n\}_{n<\omega}$ be a countable open cover of $W(\omega_\alpha)$ with no finite subcovering. Let $O_n = \bigcup_{k\leq n} U_k$. For each n choose σ_n to be the least element of $W(\omega_\alpha) \setminus O_n$. Then $\sigma_0 \leq \sigma_1 \leq \sigma_2 \leq \dots$ and infinitely many of the $\leq$ are actually $<$. Let $\sigma = \text{lub}\,\{\sigma_n\}$. There is a integer k for which $\sigma \in U_k \subset O_k$. There is another integer p for which $\sigma_n \in O_k$ if $n \geq p$. This gives an immediate contradiction for all $n \geq k$.

4.11 Corollary. W is countably compact.

4.12 Theorem. Each $f \in C^*(W(\omega_\alpha))$ has a continuous extension to $W(\omega_\alpha + 1)$ if $\alpha \geq 1$.

Proof. We will show that for each $f \in C^*(W(\omega_\alpha))$ there exist $\alpha_f \in W(\omega_\alpha)$ and $r_f \in \mathbb{R}$ such that $\sigma \geq \alpha_f$ implies $f(\sigma) = r_f$.

For each $\sigma \in W(\omega_\alpha)$ let $S(\sigma) = \{\tau \in W(\omega_\alpha) \mid \tau \geq \sigma\}$. Then $S(\sigma)$ is countably compact (same proof as 4.10) and so $\{f(S(\sigma)) \mid \sigma \in W(\omega_\alpha)\}$ is a family of countably compact hence compact hence closed subsets of $f(W(\omega_\alpha))$ (a compact space) with the finite intersection property. Let $r_f \in \bigcap_{\sigma<\omega_\alpha} f(S(\sigma))$. Let $H = f^{-1}(r_f)$ and $K_n = f^{-1}((-\infty, r_f - 1/n]) \cup f^{-1}([r_f + 1/n, \infty))$. Then, for each n, H and K_n are disjoint closed sets and, as H is unbounded, K_n is bounded, say by σ_n. Let $\alpha_f = \text{lub}\,\{\sigma_n\} + 1$. Then if $\sigma \geq \alpha_f$ we have $\sigma \notin K_n$ for any n. Thus $f(\sigma) = r_f$ if $\sigma \geq \alpha_f$. Define $f^*\colon W(\omega_\alpha + 1) \to \mathbb{R}$ by

$$f^*(\sigma) = \begin{cases} f(\sigma) & \text{if } \sigma \in W(\omega_\alpha) \\ r_f & \text{if } \sigma = \omega_\alpha \end{cases}$$

Clearly f^* is continuous and extends f.

4.13 Corollary. $\beta W(\omega_\alpha) \approx W(\omega_\alpha + 1)$.

Proof. We can consider $W(\omega_\alpha + 1)$ to be a compactification of $W(\omega_\alpha)$, using the inclusion mapping $i\colon W(\omega_\alpha) \to W(\omega_\alpha + 1)$ as an embedding onto a dense subset. 4.12 says each $f \in C^*(W(\omega_\alpha))$ has a continuous extension to $W(\omega_\alpha + 1)$; this is the property which characterizes $\beta W(\omega_\alpha)$ (2.11).

<u>4.14 Corollary</u>. $\beta W \approx W^*$.

<u>4.15 Proposition</u>. If X is compact and $|X| < \aleph_\alpha$ then $\beta(X \times W(\omega_\alpha)) \approx X \times W(\omega_\alpha + 1)$, $\alpha \geq 1$.

<u>Proof</u>. Since $X \times W(\omega_\alpha)$ is dense in $X \times W(\omega_\alpha + 1)$, a compact set, we need only to show that each $f \in C^*(X \times W(\omega_\alpha))$ has a continuous extension to $X \times W(\omega_\alpha + 1)$. For $x \in X$, let r_x be the real number to which $f|_{\{x\} \times W(\omega_\alpha)}$ is eventually equal. Define $f^*: X \times W(\omega_\alpha + 1) \to \mathbb{R}$ by

$$f^*(x,\sigma) = \begin{cases} f(x,\sigma), & \sigma < \omega_\alpha \\ r_x \quad , & \sigma = \omega_\alpha \end{cases}$$

For each x there is a $\sigma_x \in W(\omega_\alpha)$ such that $f(x,\sigma) = r_x$ if $\sigma > \sigma_x$. Let σ^* be the lub of the set $\{\sigma_x | x \in X\}$ (σ^* exists by 4.6).

To see that f^* is continuous at (x, ω_α) consider the following. For $\varepsilon > 0$ there is in $X \times W(\omega_\alpha)$ a neighborhood of $(x, \sigma^* + 1)$ of the form $0 \times U$ which f sends into the interval $(r_x - \varepsilon, r_x + \varepsilon)$ since f is continuous at $(x, \sigma^* + 1)$. It is easily seen that f^* sends $0 \times [\sigma^* + 1, \omega_\alpha]$ into $(r_x - \varepsilon, r_x + \varepsilon)$. Thus f^* is continuous at (x, ω_α).

<u>4.16 Corollary [CH]</u>. $\beta([0,1] \times W(\omega_2)) \approx [0,1] \times W(\omega_2 + 1)$.

<u>4.17 Proposition</u>. For any space X there is a space Y for which X is homeomorphic to $\beta Y \setminus Y$.

<u>Proof</u>. Choose α for which $|\beta X| < \aleph_\alpha$. By 4.15 $\beta(\beta X \times W(\omega_\alpha)) \approx \beta X \times W(\omega_\alpha + 1)$. Let $Y = [\beta X \times W(\omega_\alpha + 1)] \setminus [X \times \{\omega_\alpha\}]$. Then

$$\beta X \times W(\omega_\alpha) \subset Y \subset \beta X \times W(\omega_\alpha + 1) \quad \text{so}$$

that $\beta Y = \beta X \times W(\omega_\alpha + 1)$. [This is easily seen in general: If $S \subset T \subset \beta S$ then $\beta T = \beta S$ since any $f \in C^*(T)$ may be restricted to S and then extended to βS.]

$\beta Y \setminus Y = X \times \{\omega_\alpha\}$ which is homeomorphic to X.

<u>4.18 Corollary</u>. For a given space X there is a space Y so that X is homeomorphic to $\beta Y \setminus Y$ and $\beta Y \setminus Y$ is C^*-embedded in βY.

<u>Proof</u>. By 4.17 there is a space Z for which $\beta Z \setminus Z$ is homeomorphic to βX. If we identify βX with $\beta Z \setminus Z$ and let $Y = Z \cup (\beta X \setminus X) = \beta Z \setminus X$,

then we have $\beta Y = \beta Z$ (since $Z \subset Y \subset \beta Z$) and since βX is compact we have X C^*-embedded in βX which is C^*-embedded in $\beta Z = \beta Y$. Thus X is C^*-embedded in βY.

We have considered (above) a class of spaces X whose Stone-Čech compactification is especially simple. We go to the other extreme now and consider βN, a very complicated object. This space was first studied by Čech [1937]. We begin with a result which Čech did not obtain, although he defined the problem: what is the cardinality of βN? We need the following well-known result.

4.19 Theorem. (Pondiczery [1944], Hewitt [1946], Marczewski [1947]). The product of a family of c separable spaces is separable.

Proof. Suppose X_r is separable for each $r \in \mathbb{R}$. Let D_r be a countable dense subset of X_r and let $f_r\colon N_r \to D_r$ be an onto mapping where $N_r = N$. Define $f\colon \prod_{r\in\mathbb{R}} N_r \to \prod_{r\in\mathbb{R}} X_r$ by

$$[f(p)](r) = f_r(p(r)).$$

$$[p \in \prod N_r \text{ means } p\colon \mathbb{R} \to \bigcup_{r\in\mathbb{R}} N_r.]$$

We will show that in $\prod_{r\in\mathbb{R}} N_r$ there is a countable dense set D. Then $f(D)$ is countable in $\prod_{r\in\mathbb{R}} X_r$ and

$$\overline{f(D)} \supset f(\overline{D}) = f(\prod_{r\in\mathbb{R}} N_r) = \prod_{r\in\mathbb{R}} D_r.$$

Since $\prod_{r\in\mathbb{R}} D_r$ is dense in $\prod_{r\in\mathbb{R}} X_r$ it follows that $f(D)$ is dense in $\prod_{r\in\mathbb{R}} X_r$.

To this end, let $\mathcal{B}$ the family of all finite sets of disjoint closed intervals of $\mathbb{R}$ having rational end points. Then $\mathcal{B}$ is countable so that if

$$D = \{f\colon \mathbb{R} \to \cup N_r \mid \text{there is a } \{I_1,\dots,I_k\} \in \mathcal{B}$$

for which f is constant on each I_j and $f \equiv 1$ on $\mathbb{R} \setminus \bigcup_{j=1}^{k} I_j\}$ it follows that D is countable.

Let V be an arbitrary basic open set in $\prod_{r\in\mathbb{R}} N_r$. Then $V = \prod_{r\in\mathbb{R}} W_r$ where W_r is open in N_r and $W_r = N_r$ except for $r \in \{r_1,\dots,r_k\}$. Let $I_1,\dots,I_k$ denote pairwise disjoint closed intervals in $\mathbb{R}$ with rational endpoints and with $r_i \in I_i$, $i = 1,2,\dots,k$. Choose $n_i \in W_{r_i}$, $i = 1,2,\dots,k$, and define $f\colon \mathbb{R} \to \cup N_r$ by

$$f(r) = \begin{cases} n_i & r \in I_i,\ i = 1,2,\dots,k \\ 1 & r \in \mathbb{R} \setminus \bigcup_{i=1}^{k} I_i. \end{cases}$$

Clearly $f \in D$. We claim $f \in V$. This is easily seen from the fact that $n_i \in W_{r_i}$.

We have shown that every basic open set in $\prod_{r \in \mathbb{R}} N_r$ intersects D. Thus D is dense in $\prod_{r \in \mathbb{R}} N_r$.

4.20 Theorem. (Pospíšil [1937]). $|\beta N| = 2^c$.

Proof. Let $X = \prod_{r \in \mathbb{R}} I_r$, $I_r = [0,1]$, and let D be a countable dense subset of X with $f\colon N \to D$ (onto). On $\alpha N = N \cup X$ specify the basic open sets to be (i) $\{p\}$, if $p \in N$ and
(ii) $V \cup [f^{-1}(V) \setminus f^{-1}(p)]$, where V is any neighborhood of p in X.
It is easily verified that $\alpha N \in K(N)$. Since αN is a quotient space of βN it follows that $|\beta N| \geq |\alpha N| = 2^c$; $|\beta N| \leq 2^c$ by 3.21.

4.21 Remarks. The use of the Pondiczery-Hewitt-Marczewski theorem (4.19) is clearly restricted (above) to obtaining a compact separable space X with large cardinality. Actually, any separable space Y (which is completely regular) with cardinality 2^c could be used to obtain such a space: let $X = \beta Y$. As indicated, 4.20 is originally due to Pospíšil. However, this proof was suggested by Engelking [1968], (See also 8.7).

We next give a curious result, originally due to Sierpínski [1938] and improved by Semadeni [1964] and Dodziuk [1969]. Čech [1937] had guessed that it was "impossible to determine effectively (in the sense of Sierpínski) a point of $\beta I \setminus I$". (He used I for the countably infinite discrete space.) The following result confirms his conjecture.

4.22 Theorem. To each point of $\beta N \setminus N$ there corresponds a non-Lebesgue measurable function $f\colon \mathbb{R} \to \mathbb{R}$.

Proof. A point of $\beta N \setminus N$ can be considered to be a free z-ultrafilter on N (3.11, 3.12) and, since N is discrete, it can be considered to be a free ultrafilter on N. Let $\mathcal{F}$ be one such. For each $x \in \mathbb{R}$ let $x = k + \sum_{n=1}^{\infty} c_n/2^n$ be the normal dyadic expansion of x (i.e., $c_n = 0$ or 1 and $c_n = 0$ for infinitely many n). Define $f\colon \mathbb{R} \to \mathbb{R}$ by

$$f(x) = \begin{cases} 0 & \text{if } \{n \mid c_n = 0\} \in \mathcal{F} \\ 1 & \text{otherwise.} \end{cases}$$

Properties: (i) If x is dyadic ($c_n = 0$ for $n \geq n_o$) then $f(x) = 0$. [Each point of $\beta N \setminus N$ is a limit point of the set $A = \{n \mid c_n = 0\}$ since A contains the set $\{n \mid n \geq n_o\}$. Thus $\mathcal{F} \in \overline{A}$ so that $A \in \mathcal{F}$ (3.11).]

(ii) If A and B are subsets of $\mathbb{N}$ which are equal except for a finite set of points, then $A \in \mathcal{F}$ if and only if $B \in \mathcal{F}$. [Suppose $A \in \mathcal{F}$. Since $\mathcal{F}$ is free, any two elements of $\mathcal{F}$ intersect in an infinite set. Thus B intersects every element of $\mathcal{F}$. We conclude that $B \in \mathcal{F}$.]

(iii) If d is any dyadic number, then $f(x) = f(x + d)$ for all $x \in \mathbb{R}$. [If $x = k + \sum c_n/2^n$ and $x + d = \ell + \sum d_n/2^n$ then if $A = \{n \mid c_n = 0\}$ and $B = \{n \mid d_n = 0\}$ it follows that A and B differ in a finite set. By (ii) $A \in \mathcal{F}$ if and only if $B \in \mathcal{F}$. Thus $f(x) = f(x + d)$.]

Now suppose f is measurable. For each dyadic number b let

$$c(b) = \frac{1}{b}\int_0^b f(t)dt.$$

Claim: For b and d dyadic $c(b) = c(d)$. Since b and d are dyadic, there are integers p, q, k for which $b = p/2^k$, $d = q/2^k$.

$$c(b) = \frac{1}{b}\int_0^b f(t)dt = \frac{2^k}{p}\int_0^{p/2^k} f(t)dt = \frac{2^k \cdot p}{p}\int_0^{1/2^k} f(t)dt =$$

$$2^k\int_0^{1/2^k} f(t)dt$$

and

$$c(d) = \frac{1}{d}\int_0^d f(t)dt = \frac{2^k}{q}\int_0^{q/2^k} f(t)dt = \frac{2^k \cdot q}{q}\int_0^{1/2^k} f(t)dt =$$

$$2^k\int_0^{1/2^k} f(t)dt.$$

Thus, $c(b) = c(d)$. Let c be this constant value; i.e.,

$$c = \frac{1}{x}\int_0^x f(t)dt, \quad x \text{ dyadic}.$$

Define $g\colon \mathbb{R} \to \mathbb{R}$ by

$$g(x) = \int_0^x f(t)dt - cx.$$

Clearly g is continuous and for b dyadic we have

$$\begin{aligned} g(x + b) &= \int_0^{x+b} f(t)dt - c(x + b) \\ &= \int_0^x f(t)dt - cx + \int_x^{x+b} f(t)dt - cb \\ &= g(x) + \int_0^b f(t)dt - cb = g(x) + cb - cb \\ &= g(x). \end{aligned}$$

We conclude that $g(x)$ is constant. (It is continuous with a dense set of periods.) Thus $g(x) \equiv g(0) = 0$ and it follows that $f(x) = c$ for almost all $x \in \mathbb{R}$.

Now, for any non-dyadic number $x = \sum c_n/2^n$ in $[0,1]$ we have $1 - x = \sum 1/2^n - \sum c_n/2^n = \sum (1 - c_n)/2^n$ so that the set $A = \{n | c_n = 0\}$ has for its complement in N the set $B = \{n | 1 - c_n = 0\}$. Since F is an ultrafilter on N we have $A \in F$ if and only if $B \notin F$. We conclude that for non-dyadic x in $[0,1]$ $f(1 - x) = 1 - f(x)$. As a consequence $c = 1 - c$. (The dyadic numbers are countable and so are of measure zero. Thus there is a non-dyadic x in $[0,1]$ for which $f(x) = c = f(1 - x)$.) Therefore $c = 1/2$. However $f(x)$ is either 0 or 1 and is never $1/2$. This contradiction says that f is non-measurable.

<u>4.23 Remarks</u>. We finish our discussion of βN by giving a proof of the non-homogeneity of $\beta N \setminus N$. The problem originated (apparently informally, as it does not seem to appear in the proceedings) at the Summer Institute on Set Theoretic Topology at Madison, Wisconsin, 1955. It was solved by Rudin [1956]. We will need a series of lemmas and definitions prior to the solution.

<u>4.24 Lemma</u>. Disjoint sets in N have disjoint closures in βN.

<u>Proof</u>. If $A \cap B = \phi$ define $f\colon N \to [0,1]$ by $f(A) = \{0\}$, $f(N \setminus A) = \{1\}$. Then $F = (f^\beta)^{-1}([0, 1/3])$ and $G = (f^\beta)^{-1}([2/3, 1])$ are disjoint closed sets in βN with $A \subset F$, $B \subset G$.

<u>4.25 Lemma</u>. $\{\overline{A} | A \subset N\}$ constitutes a base for the closed sets in βN.

<u>Proof</u>. This is the way the topology for βN was defined in the alternate construction in 3.11.

<u>4.26 Lemma</u>. $\{\overline{A} | A \subset N\}$ constitutes a base for the open sets in βN.

<u>Proof</u>. Let $A \subset N$, $B = N \setminus A$. $\overline{A} \cup \overline{B} = \beta N$ and $\overline{A} \cap \overline{B} = \phi$ by 4.24 so that $\overline{A}$ is open in βN. If $p \in \beta N$ and O is a zero set neighborhood of p in βN (i.e., there is a function $f \in C(\beta N)$ and an $\varepsilon > 0$ for which $O = \{x \in \beta N |\ |f(x)| \leq \varepsilon\}$) then if we let $g = f|_N$ and $A = \{n \in N |\ |g(n)| \leq \varepsilon\}$ then $p \in \overline{A} \subset O$. Since the zero neighborhoods form a base for the open sets in a completely regular space, the lemma follows.

<u>4.27 Definition</u>. For A an infinite subset of N let $A^* = \overline{A} \setminus A$.

<u>4.28 Lemma</u>. $B^* \subset A^*$ if and only if $B \setminus A$ is finite.

Proof. If $B \setminus A = \{n_1,\ldots,n_k\}$ then $B \setminus \{n_1,\ldots,n_k\} \subset A$. Thus $\overline{B} \setminus \{n_1,\ldots,n_k\} \subset \overline{A}$ and so $B^* = \overline{B} \setminus B \subset \overline{A} \setminus A = A^*$.

Conversely, suppose $B \setminus A = \{n_1, n_2,\ldots\} = B_1$. Since $A \cap B_1 = \phi$, all ultrafilters on N which contain B_1 (i.e., $\overline{B}_1$) do not contain A and hence do not belong to $\overline{A}$. Thus, $\overline{B} \cap \overline{A} = \phi$ and so $\overline{B}_1 \setminus N \not\subset \overline{A}$. We conclude that $B^* \not\subset A^*$.

4.29 Corollary. $B^* = A^*$ if and only if $(A \cup B) \setminus (A \cap B)$ is finite; i.e., A and B agree except for a finite set.

4.30 Lemma. There are precisely c distinct sets of the form A^*.

Proof. Clearly there are no more than c since they are determined by subsets of N.

To show there are at least c let $f\colon N \to Q$ (= rationals) be a 1 - 1, onto function and for each $r \in \mathbb{R} \setminus Q$ let $\{r_n\}_{n=1}^{\infty}$ be an increasing rational sequence converging to r and let $E_r = \{f^{-1}(r_n)\}_{n=1}^{\infty}$. Then $E_r \subset N$ is infinite and $E_r \cap E_{r'}$ is finite for $r \neq r'$. Thus $E_r^* \neq E_{r'}^*$ if $r \neq r'$. There are c of these sets, one for each $r \in \mathbb{R} \setminus Q$.

4.31 Lemma. (i) $\{A^* | A \subset N\}$ is a base for the closed sets in $\beta N \setminus N$.
(ii) $\{A^* | A \subset N\}$ is a base for the open sets in $\beta N \setminus N$.

Proof. Immediate from 4.25 and 4.26.

4.32 Lemma. If $\{A_n^*\}_{n=1}^{\infty}$, $A_n^* \neq A_m^*$ if $m \neq n$ has the finite intersection property then $\bigcap_{n=1}^{\infty} A_n^*$ contains a non-empty set of the form A^*, $A \subset N$.

Proof. Choose $x \in A_1$. If distinct $x_1, x_2,\ldots,x_{n-1}$ have been chosen with $x_i \in A_1 \cap\ldots\cap A_i$ then let $B_i = A_i \setminus \{x_1,\ldots,x_{n-1}\}$. By 4.29 we have $B_i^* = A_i^*$ so that $B_n^* \cap [B_1^* \cap\ldots\cap B_{n-1}^*] \neq \phi$. By 4.24 $B_n \cap [B_1 \cap\ldots\cap B_{n-1}] \neq \phi$. Choose x_n in this intersection. Then $x_n \in A_1 \cap\ldots\cap A_n$.

Let $A = \{x_1,x_2,\ldots\}$. Clearly $A \setminus A_n$ is a finite set for each n. Thus, by 4.28 we have $A^* \subset A_n^*$ for each n.

4.33 Definition. A point $x \in X$ is a P-point if each $f \in C(X)$ is constant on a neighborhood of x.

4.34 Theorem [CH]. $\beta N \setminus N$ contains P-points.

Proof. Assuming the continuum hypothesis, we may index the family

$\{A^*|A \subset N,\ A$ infinite$\}$ with the set of countable ordinals W. Thus, we may assume this family to be denoted $\{V_\alpha\}_{\alpha<\omega_1}$. Define a family $\{U_\alpha\}_{\alpha<\omega_1}$ inductively:

$$U_o = V_o.$$

Suppose U_τ has been defined for each $\tau < \alpha < \omega_1$ so that U_τ is either some V_σ or the intersection of two of the V's. Also, $\bigcap_{\sigma\leq\tau} U_\sigma \neq \phi$.

By 4.32 there is a $V_\gamma \subset \bigcap_{\tau<\alpha} U_\tau$.

Let
$$U_\alpha = \begin{cases} V_\alpha \cap V_\gamma & \text{if } V_\alpha \cap V_\gamma \neq \phi \\ V_\gamma & \text{otherwise.} \end{cases}$$

$\{U_\alpha\}_{\alpha<\omega_1}$ is a family of closed subsets of $\beta N \setminus N$ which has the finite intersection property. (Actually, it is nested: $\sigma < \tau$ implies $U_\tau \subset U_\sigma$.) Since $\beta N \setminus N$ is compact there is a $p \in \bigcap_{\alpha<\omega_1} U_\alpha$.

If $p \in V_\alpha$ then $U_\alpha = V_\alpha \cap V_\gamma$ (from the above definition of U_α) so that $U_\alpha \subset V_\alpha$. Suppose $q \neq p$. Then there exist V_σ, V_τ such that $p \in V_\sigma$, $q \in V_\tau$ and $V_\sigma \cap V_\tau = \phi$ (4.31 (ii)). Since $U_\sigma \subset V_\sigma$ it follows that $q \notin \bigcap_{\alpha<\omega_1} U_\alpha$. Therefore, $\bigcap_{\alpha<\omega_1} U_\alpha = \{p\}$.

We claim that p is a P-point of $\beta N \setminus N$. For suppose $f \in C(\beta N \setminus N)$ and $f(p) = r$. Then $0_n = f^{-1}((r - 1/n, r + 1/n))$ is an open subset of $\beta N \setminus N$ containing p. By 4.31 (ii) there is an $\alpha_n < \omega_1$ such that $p \in V_{\alpha_n} \subset 0_n$. It follows that $p \in U_{\alpha_n} \subset V_{\alpha_n}$ and so, if $\alpha = \text{glb}\{\alpha_n\}$ we have $U_\alpha \subset \bigcap_{n=1}^{\infty} U_{\alpha_n} \subset \bigcap_{n=1}^{\infty} V_{\alpha_n} \subset \bigcap_{n=1}^{\infty} 0_n$. Thus, on U_α (a neighborhood of p) f is identically equal to r; i.e., p is a P-point of $\beta N \setminus N$.

<u>4.35 Theorem</u>. $\beta N \setminus N$ contains non P-points.

<u>Proof</u>. Suppose every point of $\beta N \setminus N$ were a P-point. If $f \in C(\beta N \setminus N)$ then for each $x \in \beta N \setminus N$ there is a neighborhood 0_x of x on which f is constant. Since $\beta N \setminus N$ is compact, there are points $x_1, \ldots, x_n$ for which $\beta N \setminus N = \bigcup_{i=1}^{n} 0_{x_i}$. This implies that f has a finite range, namely $\{f(x_1), \ldots, f(x_n)\}$. Thus, to prove the theorem it suffices to demonstrate that some function on $\beta N \setminus N$ has infinite range.

Let g: $N \to Q \cap [0,1]$ be onto. g has an extension g^β: $\beta N \to [0,1]$ and $g^\beta(\beta N)$ is a compact subset of $[0,1]$ containing

$Q \cap [0,1]$. Thus $g^\beta(\beta N) = [0,1]$. Let $f = g^\beta|_{\beta N \setminus N}$. Clearly f has infinite range.

4.36. Corollary [Rudin]. $\beta N \setminus N$ is not homogeneous.

Proof. If p is a P-point of $\beta N \setminus N$, $h(p) = q$ where $h: \beta N \setminus N \to \beta N \setminus N$ is a homeomorphism, and $f \in C(\beta N \setminus N)$ then $f \circ h \in C(\beta N \setminus N)$ and so is constant on some neighborhood U of p. Thus we see that f is constant on the neighborhood $h(U)$ of q. This proves that q is a P-point. Thus homeomorphisms of $\beta N \setminus N$ onto itself must take P-points to P-points.

4.37 Remarks. There is a proof of the non-homogeneity of $\beta N \setminus N$ due to Frolik [1967] which does not require the use of the continuum hypothesis. Why then have we included the preceeding proof by Rudin? First, for historical reasons: it was the first. Second, it demonstrates the existence of P-points in $\beta N \setminus N$ and this has been the source of much fruitful research in this area. We will now give Frolik's proof. Essentially, it shows that the number of functions at our disposal is too few for the number of points involved. We will need two classical results from general topology.

4.38 Theorem (Tychonoff [1925]). A regular, Lindelöf space is normal.

Proof. Let A and B be disjoint, closed subsets of X, a regular, Lindelöf space. For each $x \in A$ there is an open set $U_x \subset \overline{U}_x \subset X \setminus B$ with $x \in U_x$. Similarly, for each $y \in B$ there is an open set $V_y \subset \overline{V}_y \subset X \setminus A$. $\{U_x\}_{x \in A}$ is an open cover of A (which, as a closed subset of X, is Lindelöf) so there is a sequence $\{x_i\} \subset A$ with

$$A \subset \bigcup_{i=1}^{\infty} U_{x_i}.$$

Similarly, we may obtain $\{y_i\} \subset B$ with

$$B \subset \bigcup_{i=1}^{\infty} V_{y_i}.$$

Let $S_i = U_{x_i} \setminus \bigcup_{j \le i} \overline{V}_{y_j}$, $T_i = V_{y_i} \setminus \bigcup_{j \le i} \overline{U}_{x_j}$, $U = \bigcup_{i=1}^{\infty} S_i$, $V = \bigcup_{i=1}^{\infty} T_i$.

Since $\overline{V}_{y_i} \cap A = \emptyset$, $\overline{U}_{x_i} \cap B = \emptyset$ it follows that $A \subset U$, $B \subset V$. We must show that $U \cap V = \emptyset$. Now $S_i \cap V_{y_j} = \emptyset$ for $j \le i$ so that $S_i \cap T_j = \emptyset$ if $j \le i$. Also, $U_{x_j} \cap T_i = \emptyset$ for $j \le i$. Thus, $S_j \cap T_i = \emptyset$ for $j \le i$. We conclude that $S_i \cap T_j = \emptyset$ for all i,j

and hence $U \cap V = \phi$.

4.39 Theorem (Tietze [1915] - Urysohn [1925]). A closed subset of a normal space is C-embedded.

Proof. Let X be a normal space and let A be a closed subset of X. We first show that if $f(A) \subset [-1, 1]$ there is an extension. Let $H_o = f^{-1}([-1, -1/3])$, $K_o = f^{-1}([1/3, 1])$. By Urysohn's theorem [1.8] there is $g_o \in C^*(X)$ such that $g_o(X) \subset [0,1]$, $g_o(H_o) = \{0\}$, $g_o(K_o) = \{1\}$. Let $f_1(x) = \frac{2}{3}(g_o(x) - 1/2)$. It is obvious that $|f_1(x)| \leq 1/3$ for all $x \in X$ and $|f(x) - f_1(x)| \leq 2/3$ for $x \in A$.

Now suppose $f_1, f_2, \ldots, f_n$ have been defined so that

$|f_k(x)| \leq 1/3(2/3)^{k-1}$ for all $x \in X$, $k = 1,2,\ldots,n$ and

$|f(x) - \sum_{k=1}^{n} f_k(x)| \leq (2/3)^n$ for $x \in A$.

Let $H_n = [f - \sum_{k=1}^{n} f_k]^{-1}([-(2/3)^n, -1/3(2/3)^n])$

$K_n = [f - \sum_{k=1}^{n} f_k]^{-1}([1/3(2/3)^n, (2/3)^n])$.

By 1.8 there is a $g_n \in C^*(X)$ such that $g_n(X) \subset [0,1]$, $g_n(H_n) = \{0\}$, $g_n(K_n) = \{1\}$. Let $f_{n+1}(x) = (2/3)^{n+1}(g_n(x) - 1/2)$. Then, as before,

$$|f_{n+1}(x)| \leq 1/3(2/3)^n \quad \text{for all } x \in X$$

and

$$|f(x) - \sum_{k=1}^{n+1} f_k(x)| \leq (2/3)^{n+1} \quad \text{for all } x \in A.$$

Because of the first of these conditions the series $\sum_{k=1}^{\infty} f_k(x)$ is uniformly convergent and consequently defines a continuous function $F \in C^*(X)$. Because of the second of these conditions we have $F|_A = f$.

To see that A is C-embedded in X suppose that $f \in C(A)$. Let $i\colon \mathbb{R} \to [-1,1]$ be the homeomorphism defined by $i(x) = x/(1 + |x|)$. The function $i \circ f\colon A \to [-1,1]$ has an extension $F_1\colon X \to [-1,1]$. The set $K = F_1^{-1}(\{-1,1\})$ is a closed subset of X disjoint from A. There is, by 1.8, a function $g\colon X \to [0,1]$ such that $g(A) = \{1\}$ and $g(K) = \{0\}$. Let $F\colon X \to \mathbb{R}$ be defined by $F = i^{-1} \circ (F_1 \cdot g)$. Then $F \in C(X)$ and for $x \in A$

$$F(x) = i^{-1}(F_1(x)\cdot g(x)) = i^{-1}(F_1(x))$$

$$= i^{-1}(i(f(x))) = f(x).$$

Thus, F extends f to all of X.

4.40 Lemma. Any countable, discrete subspace of βN is C^*-embedded in βN.

Proof. Let D be a countable discrete subset of βN. First we show that $N \cup D$ is C^*-embedded in βN: If $f \in C^*(N \cup D)$ then $f|_N \in C^*(N)$ and so has an extension $F \in C^*(\beta N)$. Since N is dense in $N \cup D$ it follows that $F|_{N \cup D} = f$.

Finally, observe that D is C^*-embedded in $N \cup D$: $N \cup D$ is countable and hence is Lindelöf. By 4.38, $N \cup D$ is normal. D is closed in $N \cup D$ so by 4.39, D is C^*-embedded in $N \cup D$.

We conclude that D is C^*-embedded in βN.

4.41 Definition. Let $\sigma: N \to N$ be a permutation (i.e., σ is a $1 - 1$, onto mapping). Considered as a map of N into βN, σ has an extension $\overline{\sigma}: \beta N \to \beta N$ which is a homeomorphism (onto) and $\overline{\sigma}(N^*) = N^*$ where $N^* = \beta N \setminus N$ (by 1.30). Let $\sigma^* = \overline{\sigma}|_{N^*}$.

Define an equivalence relation on N^* by saying $p \sim q$ if there is a permutation $\sigma: N \to N$ for which $\sigma^*(p) = q$. The elements of $\mathcal{T} = N^*/\sim$ are called types of N^*. Let $t: N^* \to \mathcal{T}$ be the quotient mapping. For $p \in N^*$ we say that $t(p)$ is the type of p or that p is of type $t(p)$.

By 4.40 we conclude that if X is a countable discrete subset of βN then $\overline{X} = \beta X$ (closure in βN). Now βX is homeomorphic to βN so that $X^* = \overline{X} \setminus X$ is homeomorphic to N^*. For any such homeomorphism $h: \overline{X} \to \beta N$ and any $p \in X^*$, the type of p relative to X will be $t(h(p))$. We first claim that this definition is independent of h: Suppose $g: \overline{X} \to \beta N$ is another homeomorphism. Then $h \circ g^{-1}: \beta N \to \beta N$ is a homeomorphism which is induced by the permutation $\sigma = (h \circ g^{-1})|_N$. (Observe that $\sigma: N \to N$ since the points of N are the only isolated points of βN and the homeomorphism $h \circ g^{-1}$ must take isolated points of βN to isolated points of βN.) Thus
$h(p) = (h \circ g^{-1})(g(p)) = \sigma^*(g(p))$ so that $t(h(p)) = t(g(p))$.

We denote the type of p relative to X by $t(p,X)$.

4.42 Lemma (Steiner and Steiner [1971]). Suppose X and Y are countable discrete subspaces of βN.

(a) If $Y \subset X$ and $p \in X^* \cap Y^*$ then $t(p,X) = t(p,Y)$.

(b) For $p \in X^*$, $q \in Y^*$ then $t(p,X) = t(q,Y)$ if and only if there is a homeomorphism $h: \overline{X} \to \overline{Y}$ (onto) with $h(p) = q$.

(c) If h is a homeomorphism of βN onto βN or of N^* onto N^* and $p \in X^*$ then $t(p,X) = t(h(p), h(X))$.

Proof. (a) Select a homeomorphism $g: \overline{Y} \to \overline{X}$ such that $g(p) = p$. Then for any homeomorphism $h: \overline{X} \to \beta N$ we have

$$t(p,X) = t(h(p)) = t(h \circ g(p)) = t(p,Y).$$

(b) If $t(p,X) = t(q,Y)$ then there are homeomorphisms $f: \overline{X} \to \beta N$, $g: \overline{Y} \to \beta N$ with $t(f(p)) = t(g(q))$. Thus there is a homeomorphism $k: \beta N \to \beta N$ for which $k(f(p)) = g(q)$. Let $h = g^{-1} \circ k \circ f$.

Conversely, if $h: \overline{X} \to \overline{Y}$ with $h(p) = q$ and if $k: \overline{Y} \to \beta N$ is any homeomorphism, we have $t(q,Y) = t(k(q)) = t(k[h(p)]) = t(p,X)$, since $k \circ h: \overline{X} \to \beta N$ is a homeomorphism.

(c) is clear from (b).

4.43 Definition. For any infinite $S \subset \beta N$ and $p \in N^*$ define $\tau[p,S] = \{t(p,X) | X$ is a countable, discrete subset of $S\}$.

4.44 Theorem (Frolik [1967]). $|\tau[p,N^*]| \leq c$ for any $p \in N^*$.

Proof. If $\{M_n\}_{n=1}^{\infty}$ is a partition of N into infinitely many, pairwise disjoint, infinite sets and we select $y_n \in M_n^*$ for each n then p may or may not belong to the closure of the set $Y = \{y_n\}_{n=1}^{\infty}$. Considering all possible such partitions of N, let $\mathcal{Y}$ be the family of all Y selected as above where $p \in \overline{Y}$.

For $Y \in \mathcal{Y}$, let $\mathcal{X}_Y = \{X \subset Y | p \in \overline{X} \setminus Y\}$ (closure is taken in βN). For two possible choices Y_1, Y_2 from the same partition, $p \in \overline{Y}_1 \cap \overline{Y}_2$ so that $Y_1 \cap Y_2$ is an infinite set containing p in its closure. (We must prove this claim: If $Y_1 = \{y_n^1\}_{n=1}^{\infty}$ and $Y_2 = \{y_n^2\}_{n=1}^{\infty}$ we may think of y_n^1 and y_n^2 as ultrafilters on N. If $Y_1 \cap Y_2$ were finite, we could then select $R_n \in y_n^1$, $S_n \in y_n^2$ with $R_n \cup S_n = M_n$ and $R_n \cap S_n = \phi$ (for those n where $y_n^1 \neq y_n^2$). Then for $Z \in p$ we have

$$Z = (Z \cap (\cup R_n)) \cup (Z \cap (\cup S_n)).$$

Since these are complementary sets, exactly one is in p. Now $p \in \overline{Y}_1$ so $p \in \overline{\cup R_n}$ and thus $\cup R_n \in p$. It follows that $\cup S_n \notin p$ so $p \notin \overline{Y}_2$.)

Therefore, for any given partition we need choose only one such Y (to produce a given $\mathcal{X}_Y$). Let $\mathcal{X}_p = \bigcup_{Y \in \mathcal{Y}} \mathcal{X}_Y$. Since each Y contains only c infinite subsets and since N has only c partitions as above, it follows that $|\mathcal{X}_p| \leq c \cdot c = c$.

The set $\mathcal{X}_p$ has the following properties:

(i) If $X \in \mathcal{X}_p$ then $p \in \overline{X}$.

(ii) If Y is a countable discrete subset of βN with $p \in Y^*$ then there is $X \in \mathcal{X}_p$ with $X \subset Y$.

By 4.42 (a) we have $t(p,Y) = t(p,X)$. Thus $\tau[p,N^*] = \{t(p,Y) | Y$ is a countable infinite discrete subspace of βN and $p \in \overline{Y}\} \subset \{t(p,X) | X \in X_p\}$. We conclude that

$$|\tau[p,N^*]| \leq |X_p| \leq c.$$

<u>4.45 Theorem</u>. If $h\colon N^* \to N^*$ is a homeomorphism and $h(p) = q$ then $\tau[p,N^*] = \tau[q,N^*]$.

<u>Proof</u>. If X is any countably infinite, discrete subspace of N^* with $p \in \overline{X}$ then $t(p,X) = t(q, h(X))$ by 4.42 (c).

Similarly, if Y is any countably infinite, discrete subspace of N^* with $q \in \overline{Y}$ then

$$t(q,Y) = t(p, h^{-1}(Y)).$$

We conclude that $\tau[p,N^*] = \tau[q,N^*]$.

<u>4.46 Corollary</u>. N^* is not homogeneous.

<u>Proof</u>. By 4.45, a homeomorphism of N^* onto itself can take p to q only if $\tau[p,N^*] = \tau[q,N^*]$. $|\tau[p,N^*]| \leq c$ for any $p \in N^*$ so p can be taken to one of at most c points of N^* by any given homeomorphism. However, $|N^*| = 2^c$.

<u>4.47 Remarks</u>. Extending these methods, Frolik [1967a] proved that $X^* = \beta X \setminus X$ is not homogeneous if X is not pseudocompact. This had been previously proved by Isiwata [1957] assuming the continuum hypothesis. For additional information on the theory of types in βN, see the book by Comfort and Negrepontis [1974].

§5: PROPERTIES OF K(X)

5.1 Remarks. We have previously observed two general results due to Lubben [1941] concerned with properties of $K(X)$. 2.7: $K(X)$ is a complete upper semilattice. 2.19: $K(X)$ is a complete lattice if and only if X is locally compact. In this chapter we will further explore $K(X)$ beginning with two results of Flaksmaier and Visliseni [1965]. The first of these is somewhat surprising: Any infinite cardinal is the cardinal number of $K(X)$ for an appropriate X. We first prove a lemma which extends a result we had already shown for N. (We did not state explicitly the exact analogue of this lemma, but it is an immediate consequence of 4.40.)

5.2 Lemma. If D is an infinite discrete space then $\beta D \setminus D$ contains a homeomorphic copy of βD.

Proof. Partition D into disjoint infinite subsets $\{A_i\}_{i \in I}$ where $|I| = |D|$. (This can be done for infinite D since $\aleph_0 \cdot |D| = |D|$.) Since D is discrete, $\overline{A}_i$ (closure in βD) is both open and closed in βD (same proof as 4.24). Select one element $a_i \in \overline{A}_i \setminus A_i$ for each i and let $T = \{a_i | i \in I\}$. Then T is a discrete subspace of $\beta D \setminus D$ with the same cardinal number as D. Suppose $f \in C^*(T)$. Define $\hat{f}: \bigcup_{i \in I} \overline{A}_i \to \mathbb{R}$ by $\hat{f}(p) = f(a_i)$ for $p \in \overline{A}_i$. Clearly $\hat{f}$ is continuous. Let $g = \hat{f}|_D$. g has a unique extension g^β: $\beta D \to \mathbb{R}$, and it follows that g^β is an extension of f to βD. Let $f^\beta = g^\beta|_{\overline{T}}$. $\overline{T}$ is a closed subspace of βD and so $\overline{T}$ is a compactification of T with the property that each $f \in C^*(T)$ has a continuous extension f^β: $\overline{T} \to \mathbb{R}$. This says $\overline{T} = \beta T$. As $\beta D \setminus D$ is closed and contains T, $\overline{T} \subset \beta D \setminus D$. Since T and D are homeomorphic it follows that $\overline{T}$ and βD are homeomorphic.

5.3 Construction. Let $X' = \beta D \setminus \overline{T}$. Then $D \subset X' \subset \beta D$ so that $\beta X' = \beta D$. If $X = \beta D \setminus T$ then we have $X' \subset X \subset \beta D$ so that $\beta X = \beta D$ and so $\beta X = X \cup T$ or $\beta X \setminus X = T$.

5.4 Theorem. $|K(X)| = |D|$.

Proof. Suppose $\alpha X \in K(X)$. Let $f_\alpha: \beta X \to \alpha X$ be the quotient mapping (2.16). For each $p \in \alpha X \setminus X$ we have $f_\alpha^{-1}(p)$ is a closed subset of βX and hence is compact. However, $f_\alpha^{-1}(p) \in \beta X \setminus X$ by 1.30 and $\beta X \setminus X = T$, a discrete subspace of $\beta X = \beta D$. Thus $f_\alpha^{-1}(p)$ is finite.

We claim $\{f_\alpha^{-1}(p) | p \in \alpha X \setminus X$ and $|f_\alpha^{-1}(p)| \geq 2\}$ is finite. Suppose not. Then we may select sets $A = \{a_i\}_{i=1}^{\infty} \subset \beta X \setminus X = T$ and $B = \{b_i\}_{i=1}^{\infty} \subset \beta X \setminus X = T$ so that $a_i \neq b_i$ but $f_\alpha(a_i) = f_\alpha(b_i)$ for all i. Since $A \subset T$, $B \subset T$, and T is discrete, we have $\overline{A} \cap \overline{B} = \phi$ (closures in βT) and because βT is equal to $\overline{T}$ (closures in βD), it follows that $\overline{A} \cap \overline{B} = \phi$ (closures in βD). Let $a \in \overline{A} \setminus A$. Then $a \in X$ so that $f_\alpha^{-1} f_\alpha(a) = a \in \beta X \setminus \overline{B}$. $f_\alpha(\overline{B})$ is compact and so is closed in αX, and $f_\alpha(a) \notin f_\alpha(\overline{B})$. Thus, there is an open U in αX containing $f_\alpha(a)$ and contained in $\alpha X \setminus f_\alpha(\overline{B})$. Thus, $f_\alpha^{-1}(U) \cap \overline{B} = \phi$. Now $a \in \overline{A}$ so that $f_\alpha(a) \in f_\alpha(\overline{A}) \subset \overline{f_\alpha(A)} = \overline{f_\alpha(B)}$ and U is a neighborhood of $f_\alpha(a)$. Thus, $f_\alpha(B) \cap U \neq \phi$. We have a $b_i \in B$ for which $f_\alpha(b_i) \in U$. This says that $f_\alpha^{-1}(f_\alpha(b_i)) \subset f_\alpha^{-1}(U)$; or $b_i \in f_\alpha^{-1}(U)$ contradicting the fact that $f_\alpha^{-1}(U) \cap \overline{B} = \phi$.

We have shown that for a member αX of K(X), the set $\{f_\alpha^{-1}(p) | p \in \alpha X \setminus X\}$ is composed entirely of singletons except for finitely many finite (non-singleton) elements. Since this set entirely determines αX (2.16), the cardinality of K(X) cannot exceed the cardinal number of the collection of such families. The collection of such families is a subset of $P_F(P_F(\beta X \setminus X))$ where $P_F(W) = \{Y \subset W | Y$ is finite$\}$. Clearly $|P_F(P_F(\beta X \setminus X))| = |\beta X \setminus X| = |D|$. Thus, $|K(X)| \leq |D|$.

To obtain $|K(X)| \geq |D|$ observe that K(X) has distinct compactifications, one for each pair $\{p,q\} \subset \beta X \setminus X$, obtained by identifying p and q. Thus, there are at least as many elements in K(X) as there are pairs in $\beta X \setminus X = D$; i.e., $|K(X)| \geq |D| \cdot |D| = |D|$.

5.5 Remark. The above proof carries the seeds of the proof of the next result, which gives a sufficient condition for K(X) to be a lattice (not necessarily complete).

5.6 Theorem. If $\beta X \setminus X$ is discrete and $\overline{\beta X \setminus X} = \beta(\beta X \setminus X)$ then K(X) is a lattice.

Proof. For any $\alpha X \in K(X)$ the above proof shows that $\{f_\alpha^{-1}(p) | p \in \alpha X\}$ is a family of subsets of βX all of which are finite and all but finitely many are singletons.

To obtain the greatest lower bound for a pair $\alpha X, \gamma X \in K(X)$ consider the subset C of $C^*(X)$ consisting of all functions g whose extension g^β to βX is constant on all non-singletons of the two families $\{f_\alpha^{-1}(p) | p \in \alpha X\}$, $\{f_\gamma^{-1}(q) | q \in \gamma X\}$ and subject to the condition that whenever $f_\alpha^{-1}(p) \cap f_\gamma^{-1}(q) \neq \phi$ $(p \in \alpha X, q \in \gamma X)$ then $g^\beta(f_\alpha^{-1}(p)) = g^\beta(f_\gamma^{-1}(q))$.

C separates points and closed sets in X: Suppose $F = K \cap X$, K closed in βX and $x \in X \setminus F$. Let K_1 be the union of K with all non-singletons in the two families $\{f_\alpha^{-1}(p) | p \in \alpha X\}$, $\{f_\gamma^{-1}(q) | q \in \gamma X\}$. Then K_1 is closed in βX and $x \notin K_1$. Let $f^\beta(x) = 0$, $f^\beta(K_1) = \{1\}$. Then $f \in C$ and separates x and F.

Thus C determines a compactification $e_C X$ (2.4) and since $C \subset C_\alpha \cap C_\gamma$ it follows that $\{\delta X \in K(X) | \delta X \leq \alpha X, \delta X \leq \gamma X\}$ is non-empty. This family has a lub εX. Clearly $\varepsilon X = \alpha X \wedge \gamma X$.

5.7 Remarks. Another result of Visliseni and Flaksmaier is a sufficient condition for $K(X)$ to fail to be a lattice. Its proof requires an analogue of Urysohn's theorem (1.8) for completely regular spaces. Of course, one cannot expect to get quite as good a conclusion with the hypothesis of normality weakened.

5.8 Lemma. If F and G are disjoint closed subsets of X and F is compact then there is an element $f \in C^*(X)$ for which $f(F) = \{0\}$ and $f(G) = \{1\}$.

Proof. For each $x \in F$ there is an element $f_x\colon X \to [0,1]$ such that $f_x(x) = 0$, $f_x(G) = \{1\}$. $\{f_x^{-1}([0, 1/2))\}_{x \in F}$ is an open cover of the compact set F so that there is a finite collection $\{x_1, \ldots, x_n\}$ for which $\bigcup_{i=1}^n f_{x_i}^{-1}([0, 1/2)) \supset F$. Let $g \in C^*(X)$ be defined by $g = f_{x_1} \cdot f_{x_2} \ldots f_{x_n}$. Then $g\colon X \to [0,1]$, $g(F) \subset [0, 1/2)$ (since for each point of F, one of the f_{x_i} is less than $1/2$ there and all the other functions are no larger than 1 there), and $g(G) = \{1\}$.

If $h\colon \mathbb{R} \to \mathbb{R}$ is defined by

$$h(r) = \begin{cases} 0 & \text{if } r \leq 1/2 \\ 2r - 1 & \text{if } 1/2 \leq r \leq 1 \\ 1 & \text{if } r \geq 1. \end{cases}$$

then $f = h \circ g$ is the desired function.

5.9 Theorem. If a sequence in $\beta X \setminus X$ converges to a point of X, then $K(X)$ is not a lattice.

Proof. Suppose $\{x_i\}_{i=1}^{\infty} \subset \beta X \setminus X$ and $\lim_{i\to\infty} x_i = x \in X$. Let
$\mathcal{F} = \{f \in C^*(X) | f^\beta(x_{2i}) = f^\beta(x_{2i-1})$ for all $i\}$ and let
$\mathcal{G} = \{f \in C^*(X) | f^\beta(x_{2i}) = f^\beta(x_{2i+1})$ for all $i\}$.

We claim that the families $\mathcal{F}$ and $\mathcal{G}$ both separate points from closed sets in X. For suppose F is closed in X and $p \in X \setminus F$. The only case which causes difficulty is if $p = x$. Then $F = K \cap X$, where K is closed in βX. K does not contain x so that K can contain at most finitely many of the x_i. We can, if necessary, add finitely many of the x_i to K so that K contains $x_1,\ldots,x_{2k}$ for some k. K and $\{x\} \cup \{x_i\}_{i=2k+1}^{\infty}$ are disjoint closed sets in βX. Let f^β be a function in $C^*(\beta X)$ for which $f^\beta(K) = \{0\}$ and $f^\beta(\{x\} \cup \{x_i\}_{i=2k+1}^{\infty}) = \{1\}$ (5.8). Then f separates F and x and $f \in \mathcal{F}$. ($f = f^\beta|_X$.)

The proof for $\mathcal{G}$ is similar.

Next, we assert that no element of $K(X)$ is less then $e_{\mathcal{F}}X$ and $e_{\mathcal{G}}X$. For suppose there is an element $\alpha X \in K(X)$ and functions $f\colon e_{\mathcal{F}}X \to \alpha X$, $g\colon e_{\mathcal{G}}X \to \alpha X$ which extend the identity mappings of X in $e_{\mathcal{F}}X$, $e_{\mathcal{G}}X$ onto X in αX. Let $\pi_{\mathcal{F}}\colon \beta X \to e_{\mathcal{F}}X$, $\pi_{\mathcal{G}}\colon \beta X \to e_{\mathcal{G}}X$, $\pi_\alpha\colon \beta X \to \alpha X$ be the quotient mappings. The following diagram commutes.

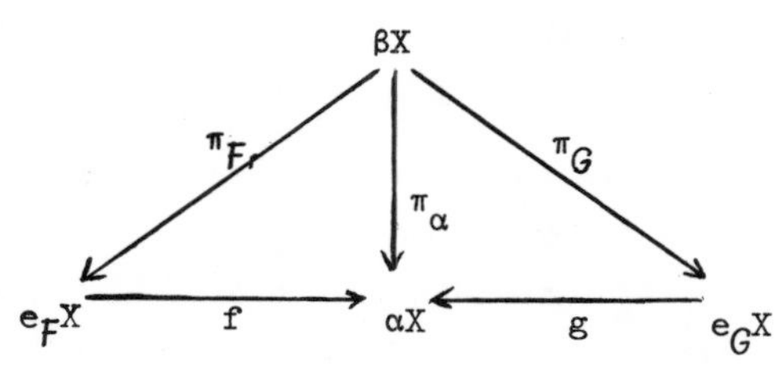

By 1.30 $\pi_\alpha(\beta X \setminus X) \subset \alpha X \setminus X$. However $\pi_\alpha(x_1) = f \circ \pi_{\mathcal{F}}(x_1) = f \circ \pi_{\mathcal{F}}(x_2) = \pi_\alpha(x_2) = g \circ \pi_{\mathcal{G}}(x_2) = g \circ \pi_{\mathcal{G}}(x_3) = \pi_\alpha(x_3) = \ldots = \pi_\alpha(x_n) = \ldots$
Thus $\{\pi_\alpha(x_n)\}_{n=1}^{\infty}$ is a constant sequence, which by continuity, converges to $\pi_\alpha(x)$. This contradicts the fact that $\pi_\alpha(x_n) \in \alpha X \setminus X$ whereas $\pi_\alpha(x) \in X$.

This contradiction assures the non-existence of such an αX.

5.10 Remarks. We need to show that the condition of 5.9 is actually possible. We use the following result of Čech [1937].

5.11 Theorem. If $x \in X$ has a countable base of neighborhoods in X then x has a countable base of neighborhoods in βX.

Proof. Since βX is completely regular, x has a base of closed zero set neighborhoods in βX. Let $U = \{p \in \beta X | \, |f(p)| \leq r\}$ be one such. Then $U \cap X$ is a neighborhood of x and so there is a $U_n \subset U \cap X$, $x \in U_n$, where $\{U_n\}_{n=1}^{\infty}$ is a neighborhood base at x in X. Then $\overline{U}_n \subset U$ (closure in βX). We have shown that $\{\overline{U}_n\}_{n=1}^{\infty}$ is a neighborhood base at x in βX.

5.12 Theorem (Shirota [1950]). If X is first countable and not locally compact then $K(X)$ is not a lattice.

Proof. $\beta X \setminus X$ is not closed in βX so some $x \in X$ is a limit point of $\beta X \setminus X$. Since x has a countable neighborhood base in βX, there is a sequence in $\beta X \setminus X$ converging to x. The theorem is thus a corollary of 5.9.

5.13 Corollary. $K(Q)$ is not a lattice.

5.14 Remarks. We will next be involved with determining to what extent βX determines $K(X)$ and vice versa. The first major results in this area are due to Magill [1968].

5.15 Definition. Let $\alpha X \in K(X)$ and let f_α: $\beta X \to \alpha X$ be the quotient mapping. The set $\{f_\alpha^{-1}(p) | p \in \alpha X \setminus X\}$ is denoted by $F(\alpha X)$ and is called the β-family of αX.

5.16 Lemma. If $\alpha X, \gamma X \in K(X)$ then $\alpha X \leq \gamma X$ if and only if each element of $F(\gamma X)$ is a subset of some element of $F(\alpha X)$.

Proof. Let h: $\gamma X \to \alpha X$ be such that h restricted to $X \subset \gamma X$ is the identity mapping onto $X \subset \alpha X$ Then $h o f_\gamma = f_\alpha$ and if $f_\gamma^{-1}(p) \in F(\gamma X)$ we have for any $q \in f_\gamma^{-1}(p)$ that $h(p) = h \circ f_\gamma(q) = f_\alpha(q)$. That is, $q \in f_\alpha^{-1}(h(p))$. We have shown that $f_\gamma^{-1}(p) \subset f_\alpha^{-1}(h(p))$, a member of $F(\alpha X)$.

Conversely, suppose each element of $F(\gamma X)$ is a subset of an element of $F(\alpha X)$. For each $p \in \gamma X \setminus X$ there is, by hypothesis, a point $q_p \in \alpha X \setminus X$ for which $f_\gamma^{-1}(p) \subset f_\alpha^{-1}(q_p)$ and q_p is uniquely determined by p since the elements of $F(\alpha X)$ are pairwise disjoint. Define h: $\gamma X \to \alpha X$ by

$$h(t) = \begin{cases} t & \text{if } t \in X \\ q_p & \text{if } t = p. \end{cases}$$

For any $r \in \beta X \setminus X$ there is a $t \in \alpha X \setminus X$ such that $f_\gamma^{-1}(f_\gamma(r)) \subset f_\alpha^{-1}(t)$

and so $h(f_\gamma(r)) = t$. Also, $f_\alpha(r) = t$ since $r \in f_\gamma^{-1}(f_\gamma(r)) \subset f_\alpha^{-1}(t)$. Thus, $h \circ f_\gamma = f_\alpha$. (We have shown that this holds on $\beta X \setminus X$. Trivially it holds on $X \subset \beta X$.)

For any closed set $F \subset \alpha X$ we have $f_\alpha^{-1}(F)$ is closed in βX. Thus $f_\alpha^{-1}(F)$ is compact in βX. From this it follows that $f_\gamma(f_\alpha^{-1}(F))$ is compact and hence closed in γX. $h \circ f_\gamma = f_\alpha$ implies that $f_\alpha^{-1}(F) = (h \circ f_\gamma)^{-1}(F) = f_\gamma^{-1} \circ h^{-1}(F)$ so that $h^{-1}(F) = f_\gamma(f_\alpha^{-1}(F))$ which is closed. Thus, h is continuous and we conclude that $\alpha X \leq \gamma X$.

<u>5.17 Corollary</u>. $\alpha X \approx \gamma X$ if and only if $\mathcal{F}(\alpha X) = \mathcal{F}(\gamma X)$.

<u>Proof</u>. $\alpha X \approx \gamma X$ if and only if $\alpha X \leq \gamma X$ and $\gamma X \leq \alpha X$ (2.3) if and only if each element of $\mathcal{F}(\alpha X)$ is contained in an element of $\mathcal{F}(\gamma X)$ and each element of $\mathcal{F}(\gamma X)$ is contained in an element of $\mathcal{F}(\alpha X)$ if and only if $\mathcal{F}(\alpha X) = \mathcal{F}(\gamma X)$ since the elements of $\mathcal{F}(\alpha X)$ are pairwise disjoint (and so are the elements of $\mathcal{F}(\gamma X)$).

<u>5.18 Lemma</u>. Let $F_1,\ldots,F_n$ be pairwise disjoint closed subsets of αX which are contained in $\alpha X \setminus X$. Let $\mathcal{F} = \{f \in C_\alpha \mid f^\alpha|_{F_i}$ is constant for each i, $1 \leq i \leq n\}$. Then $\mathcal{F}$ separates points from closed sets in X and hence determines an element of $K(X)$ which we denote $\alpha(X; F_1,\ldots,F_n)$.

<u>Proof</u>. Let F be closed in X and suppose $x \in X \setminus F$. Then $F = G \cap X$ where G is closed in αX. If $K = G \cup F_1 \cup \ldots \cup F_n$ then K is a closed set in αX and $x \notin K$. There is an element $\hat{f} \in C^*(\alpha X)$ such that $\hat{f}(x) = 0$ and $\hat{f}(K) = \{1\}$. Let $f = \hat{f}|_X$. Clearly, $f \in \mathcal{F}$, $f(x) = 0$ and $f(F) = \{1\}$. The rest of the proof follows from 2.4.

<u>5.19 Lemma</u>. Let K_1 and K_2 be non-empty closed subsets of βX contained in $\beta X \setminus X$. Then

$$\beta(X; K_1) \wedge \beta(X; K_2) = \begin{cases} \beta(X; K_1, K_2) & \text{if } K_1 \cap K_2 = \emptyset \\ \beta(X; K_1 \cup K_2) & \text{if } K_1 \cap K_2 \neq \emptyset \end{cases}$$

<u>Proof</u>. In either case, $\beta(X; K_i) \leq \beta(X; K_1, K_2)$ and $\beta(X; K_i) \leq \beta(X; K_1 \cup K_2)$, $i = 1,2$ by 5.16. Suppose $\gamma X \in K(X)$ and $\beta(X; K_1) \geq \gamma X$, $\beta(X; K_2) \geq \gamma X$. By 5.16 K_1 and K_2 are subsets of elements of $\mathcal{F}(\gamma X)$. In the first case applying 5.16 again we conclude that $\beta(X; K_1, K_2) \geq \gamma X$. Thus, $\beta(X; K_1, K_2) = \beta(X; K_1) \wedge \beta(X; K_2)$.

In the second case, since the elements of $F(\gamma X)$ are pairwise disjoint, we conclude that the elements of $F(\gamma X)$ containing K_1 and K_2 must intersect and hence must coincide. Thus, by 5.16 $\beta(X; K_1 \cup K_2) \geq \gamma X$ and we conclude that $\beta(X; K_1) \wedge \beta(X; K_2) = \beta(X; K_1 \cup K_2)$.

<u>5.20 Lemma</u>. Let $\alpha X \in K(X)$ where X is locally compact and suppose K_1, $K_2 \in F(\alpha X)$. Let $G = \{f \in C^*(X) | f^\beta|_F$ is constant for all $F \in F(\alpha X)$ and $f^\beta|_{K_1} = f^\beta|_{K_2}\}$. Then G separates points from closed sets in X and thus determines a compactification $e_G X$. $F(e_G X) = (F(\alpha X) \setminus \{K_1, K_2\}) \cup \{K_1 \cup K_2\}$.

<u>Proof</u>. If $x \in X \setminus H$ where H is a closed set in X then $H = H_1 \cap X$ where H_1 is closed in βX. Let $K = H_1 \cup \beta X \setminus X$. Then K is closed in βX and $x \notin K$. Thus there is an $f \in C^*(X)$ for which $f^\beta(x) = 0$, $f^\beta(K) = \{1\}$. It is clear that $f \in G$ and separates x and H.

The quotient map $f\colon \beta X \to e_G X$ can be viewed as a projection map (proof of 2.10) and since the elements of G do not distinguish between the points of the members of $(F(\alpha X) \setminus \{K_1, K_2\}) \cup \{K_1 \cup K_2\}$ it follows that $F(e_G X) = (F(\alpha X) \setminus \{K_1, K_2\}) \cup \{K_1 \cup K_2\}$.

<u>5.21 Lemma</u>. Let $\alpha X \in K(X)$ where X is locally compact and suppose $K_1, K_2 \in F(\alpha X)$. If H_1, H_2 are closed subsets of βX with $H_i \subset K_i$ then

$$\alpha X \wedge \beta(X; H_1 \cup H_2) = e_G X \quad (G \text{ as in } 5.20).$$

<u>Proof</u>. By 5.16 it is apparent that $e_G X \leq \alpha X \wedge \beta(X; H_1 \cup H_2)$. Suppose γX is less than or equal to both αX and $\beta(X; H_1 \cup H_2)$. Then each element of $F(\alpha X)$ is contained in a member of $F(\gamma X)$ and $H_1 \cup H_2$ is in some member of $F(\gamma X)$. Thus $K_1 \subset F_1$, $K_2 \subset F_2$, $H_1 \cup H_2 \subset F_3$ for some $F_1, F_2, F_3 \in F(\gamma X)$. $H_i \subset K_i$, $i = 1, 2$, says that $K_i \cap F_3 \neq \emptyset$, $i = 1, 2$. This implies that $F_i \cap F_3 \neq \emptyset$, $i = 1, 2$. Therefore $F_3 = F_1 = F_2$ and so each member of $F(e_G X)$ is contained in a member of $F(\gamma X)$. Thus $e_G X \geq \gamma X$ and the lemma follows.

<u>5.22 Definition</u>. An element $\alpha X \in K(X)$ is a <u>dual point</u> of $K(X)$ provided $\alpha X < \beta X$ and there does not exist $\gamma X \in K(X)$ for which $\alpha X < \gamma X < \beta X$.

5.23 Lemma. αX is a dual point of $K(X)$ if and only if there exist distinct points $p, q \in \beta X \setminus X$ with $\alpha X \approx \beta(X; \{p,q\})$.

Proof. Clearly $\beta(X; \{p,q\})$ is a dual point.

If αX is a dual point for which $F(\alpha X)$ is not a collection of all singleton sets but one which is a doubleton then either $F(\alpha X)$ contains (at least) two sets which are non-singletons or it contains a set with at least three elements. In the first case, choose p from one non-singleton and choose q from another. In the second case, choose p and q from the set with at least three elements. Then in either case,

$$\alpha X < \beta(X; \{p,q\}) < \beta X.$$

5.24 Lemma. If $\alpha X \in K(X)$ then $F(\alpha X)$ has precisely one non-singleton if and only if $\alpha X \neq \beta X$ and there do not exist distinct $\gamma_1 X, \gamma_2 X \in K(X)$ such that

(i) $\gamma_1 X, \gamma_2 X$ are dual points of $K(X)$

(ii) $\alpha X \wedge \gamma_1 X = \alpha X \wedge \gamma_2 X \neq \alpha X$

(iii) $\gamma_1 X$ and $\gamma_2 X$ are the only dual points of $K(X)$ greater than $\gamma_1 X \wedge \gamma_2 X$.

Proof. Suppose $F(\alpha X)$ has non-singletons H and K. Then there are points $a, b \in H$, $c, d \in K$. Let $\gamma_1 X = \beta(X; \{a,c\})$, $\gamma_2 X = \beta(X; \{b,d\})$. By 5.21 $F(\alpha X \wedge \gamma_1 X) = (F(\alpha X) \setminus \{H,K\}) \cup \{H \cup K\} = F(\alpha X \wedge \gamma_2 X)$. By 5.17, $\alpha X \wedge \gamma_1 X = \alpha X \wedge \gamma_2 X \neq \alpha X$. By 5.23 $\gamma_1 X$ and $\gamma_2 X$ are both dual points of $K(X)$. Suppose $\gamma X = \beta(X; \{x,y\})$ is a dual point of $K(X)$ greater than $\gamma_1 X \wedge \gamma_2 X$. Since $\gamma_1 X \wedge \gamma_2 X = \beta(X; \{a,c\}, \{b,d\})$ by 5.19 it follows that $\{x,y\} \subset \{a,c\}$ or $\{x,y\} \subset \{b,d\}$. We would have then that $\gamma X = \gamma_1 X$ or $\gamma X = \gamma_2 X$.

Thus, the existence of two non-singletons in $F(\alpha X)$ implies the existence of compactifications $\gamma_1 X, \gamma_2 X$ satisfying the conditions in (i), (ii), (iii).

Conversely, suppose $F(\alpha X)$ contains exactly one non-singleton K and suppose there are compactifications $\gamma_1 X, \gamma_2 X$ satisfying (i) and (ii). Then $\gamma_1 X = \beta(X; \{a,b\})$ and $\gamma_2 X = \beta(X; \{c,d\})$ for some $a, b, c, d \in \beta X \setminus X$. Because $\alpha X \wedge \gamma_1 X \neq \alpha X$ it follows that not both $a, b \in K$. Similarly not both $c, d \in K$. By symmetry arguments we can reduce the cases to consider to three:

Case 1: $\{a,b,c,d\} \cap K = \emptyset$.

Case 2: $\{a,b,c\} \cap K = \emptyset$, $d \in K$.

Case 3: $\{a,c\} \cap K = \emptyset$, $\{b,d\} \subset K$.

In Case 1, we have, since $\alpha X = \beta(X; K)$, $\alpha X \wedge \gamma_1 X = \beta(X; K, \{a,b\})$ by 5.19. Similarly, $\alpha X \wedge \gamma_2 X = \beta(X; K, \{c,d\})$. By (ii), $\beta(X; K, \{a,b\}) = \beta(X; K, \{c,d\})$ and so $\{a,b\} = \{c,d\}$. This says that $\gamma_1 X = \gamma_2 X$.

In Case 2, $\alpha X \wedge \gamma_1 X = \beta(X; K, \{a,b\})$ whereas $\alpha X \wedge \gamma_2 X = \beta(X; K, \{c\})$, a contradiction of (ii).

In Case 3, $\alpha X \wedge \gamma_1 X = \beta(X; K, \{a\})$ and $\alpha X \wedge \gamma_2 X = \beta(X; K, \{c\})$. By (ii) we must have $a = c$. Thus $\gamma_1 X \wedge \gamma_2 X = \beta(X; \{a,b,d\})$ and so condition (iii) is not satisfied: $\beta(X; \{b,d\})$ is a dual point above $\gamma_1 X \wedge \gamma_2 X$.

5.25 Lemma. Let X be locally compact and suppose $\alpha X \in K(X)$. Let H be a closed non-singleton subset of $\beta X \setminus X$. $H \in F(\alpha X)$ if and only if $\beta(X; H) \geq \alpha X$ and there does not exist a closed subset K of $\beta X \setminus X$ with $\beta(X; H) > \beta(X; K) \geq \alpha X$.

Proof. If $H \in F(\alpha X)$ then $\beta(X; H) \geq \alpha X$ by 5.16. No proper subset of H can belong to $F(\alpha X)$. Thus, no K exists for which $\beta(X; H) > \beta(X; K) \geq \alpha X$. If $H \notin F(\alpha X)$ then if $\beta(X; H) \geq \alpha X$ there is by 5.16 an element $K \in F(\alpha X)$ which contains H as a proper subset. Then $\beta(X; H) > \beta(X; K) \geq \alpha X$.

5.26 Theorem [Magill]. Suppose X and Y are locally compact and Γ: $K(X) \to K(Y)$ is a lattice isomorphism (onto). Then there is a homeomorphism h: $\beta X \setminus X \to \beta Y \setminus Y$ (onto) such that if $\Gamma(\alpha X) = \gamma Y$ then $F(\gamma Y) = \{h(F) | F \in F(\alpha X)\}$.

Proof. The situation is trivial if $\beta X \setminus X$ has fewer than three points. Thus, suppose $p \in \beta X \setminus X$ and $q, r \in \beta X \setminus X$ are other points so that p, q, r are distinct. $\Gamma(\beta(X; \{p,q\}))$ and $\Gamma(\beta(X; \{p,r\}))$ are dual points of $K(Y)$ so that there are points $a, b, c, d \in \beta Y \setminus Y$ for which $\Gamma(\beta(X; \{p,q\})) = \beta(Y; \{a,b\})$ and $\Gamma(\beta(X; \{p,r\})) = \beta(Y; \{c,d\})$. By 5.19 $\beta(X; \{p,q\}) \wedge \beta(X; \{p,r\}) = \beta(X; \{p,q,r\})$ so that $\Gamma(\beta(X; \{p,q,r\})) = \beta(Y; \{a,b\}) \wedge \beta(Y; \{c,d\})$. If $\{a,b\} \cap \{c,d\} = \emptyset$ then by 5.19 $\beta(Y; \{a,b\}) \wedge \beta(Y; \{c,d\}) = \beta(Y; \{a,b\}, \{c,d\})$. This is not possible since $\beta(X; \{p,q,r\})$ has exactly three dual points greater than it whereas $\beta(Y; \{a,b\}, \{c,d\})$ has exactly two dual points greater than it. Thus, $\{a,b\} \cap \{c,d\} \neq \emptyset$. As $\{a,b\} \neq \{c,d\}$, there is precisely one point common to these sets; say $a = c$. We define $h(p) = a$.

We must show that $h(p)$ is well-defined; that is, a does not depend on the choice of q and r. Suppose s is distinct from p, q, r. Then $\Gamma(\beta(X; \{p,s\})) = \beta(Y; \{x,y\})$ and we may conclude as above that $\{x,y\}$ has precisely one point in common with $\{a,b\}$ and with $\{a,d\}$. If $a \notin \{x,y\}$ then $\{x,y\} = \{b,d\}$. By 5.19 $\beta(X; \{p,q\}) \wedge \beta(X; \{p,r\}) \wedge \beta(X; \{p,s\}) = \beta(X; \{p,q,r,s\})$, $\beta(Y; \{a,b\}) \wedge \beta(Y; \{a,d\}) \wedge \beta(Y; \{b,d\}) = \beta(Y; \{a,b,d\})$. This is a contradiction since $\beta(X; \{p,q,r,s\})$ has six dual points greater than it and $\beta(Y; \{a,b,d\})$ has three dual points greater than it. We conclude that $a \in \{x,y\}$, say $a = x$, so that $\Gamma(\beta(X; \{p,s\}) = \beta(Y; \{a,y\})$. Thus, if s had been used in place of either q or r we would still have $h(p) = a$.

Now suppose H is a closed subset of $\beta X \setminus X$ having more than one point. Then $\Gamma(\beta(X; H)) = \beta(Y; K)$ for some closed subset $K \subset \beta Y \setminus Y$ (by 5.25) and K must contain more than one point. If p and q are distinct points of H then $\beta(X; \{p,q\}) \geq \beta(X; H)$ so that $\Gamma(\beta(X; \{p,q\})) = \beta(Y; \{a,b\}) \geq \Gamma(\beta(X; H)) = \beta(Y; K)$ so that $\{a,b\} \subset K$. But $h(p) \in \{a,b\}$ as is apparent from the definition of h. Thus, $h(p) \in K$. Since p was an arbitrary point of H, it follows that $h(H) \subset K$.

We may define $k\colon \beta Y \setminus Y \to \beta X \setminus X$ in an analogous manner so that if $\Gamma^{-1}(\beta(Y; \{a,b\}) = \beta(X; \{p,q\})$ then $k(a) \in \{p,q\}$. We want to show that $k \circ h$ is the identity mapping of $\beta X \setminus X$ onto $\beta X \setminus X$. If $h(p) = a$ as above and $k(a) \in \{p,q\}$ and if $k(a) \neq p$ then $k(a) = q$. If $r \neq p, q$ then there is a point $c \in \beta Y \setminus Y$ so that $\Gamma(\beta(X; \{p,r\}) = \beta(Y; \{a,c\})$. Thus $k(a) \in \{p,r\}$ and if $k(a) \neq p$ then $k(a) = r$, contradicting the fact that $k(a) = q \neq r$.

We conclude that $k \circ h = id_{\beta X \setminus X}$. Similarly, $h \circ k = id_{\beta Y \setminus Y}$. Also $h(H) \subset K$ and $k(K) \subset H$. This says that $h(H) = K$, h is $1 - 1$ and onto. Since, for an arbitrary closed set H we have $h(H)$ is closed, it follows that h is a homeomorphism.

Suppose now that $\Gamma(\alpha X) = \gamma Y$. If $H \in F(\alpha X)$ and H has more than one point then $\beta(X; H) \geq \alpha X$ so that $\beta(Y; h(H)) \geq \gamma Y$. By 5.25 there is no $V \subset \beta X \setminus X$ for which $\beta(X; H) > \beta(X; V) \geq \alpha X$. This says that there is no $W \subset \beta Y \setminus Y$ for which $\beta(Y; h(H)) > \beta(Y; W) \geq \gamma Y$.

By 5.25, $h(H) \in F(\gamma Y)$. Analogously, if $K \in F(\gamma Y)$ has more than one point, then $k(K) \in F(\alpha X)$. For $\{p\} \in F(\alpha X)$ if $\{h(p)\} \notin F(\gamma Y)$ then there is a non-singleton $K \in F(\gamma Y)$ such that $h(p) \in K$. We conclude that $p = k \circ h(p) \in k(K)$. This says that $p = k(K)$ since $\{p\} \in F(\alpha X)$.

But if K is a non-singleton, so is $k(K)$ since k is $1-1$. We conclude that $h(H) \varepsilon F(\gamma Y)$ for all $H \varepsilon F(\alpha X)$. Similarly $k(K) \varepsilon F(\alpha X)$ for all $K \varepsilon F(\gamma Y)$. Thus $F(\gamma Y) = \{h(H) | H \varepsilon F(\alpha X)\}$ and we are finished.

5.27 Theorem [Magill]. Suppose X and Y are locally compact and $h: \beta X \setminus X \to \beta Y \setminus Y$ is a homeomorphism (onto). For each $\alpha X \varepsilon K(X)$ there is a unique $\gamma Y \varepsilon K(Y)$ such that $F(\gamma Y) = \{h(F) | F \varepsilon F(\alpha X)\}$ and such that the map Γ: $K(X) \to K(Y)$ defined by $\Gamma(\alpha X) = \gamma Y$ is a lattice isomorphism.

Proof. Let $G = \{f \varepsilon C^*(Y) | f^\beta|_{h(F)}$ is constant for each $F \varepsilon F(\alpha X)\}$. Then G separates points and closed sets in Y (same proof as in 5.20) and so determines a compactification $e_G Y \varepsilon K(Y)$. Let $\gamma Y = e_G Y$. It is trivial to verify that $F(\gamma Y) = \{h(F) | F \varepsilon F(\alpha X)\}$.

If we define Γ: $K(X) \to K(Y)$ by $\Gamma(\alpha X) = \gamma Y$ then for $\alpha_1 X \neq \alpha_2 X$ we have $F(\alpha_1 X) \neq F(\alpha_2 X)$ so that $\{h(F) | F \varepsilon F(\alpha_1 X)\} \neq \{h(F) | F \varepsilon F(\alpha_2 X)\}$ and therefore $\Gamma(\alpha_1 X) \neq \Gamma(\alpha_2 X)$. Thus Γ is $1-1$.

If $\delta Y \varepsilon K(Y)$ then use h^{-1} (as we used h above) to obtain a $\varepsilon X \varepsilon K(X)$ for which $\Gamma(\varepsilon X) = \delta Y$. It follows that Γ is onto.

Now $\alpha X \leq \varepsilon X$ if and only if each member of $F(\varepsilon X)$ is contained in some member of $F(\alpha X)$. This happens if and only if each member of $\{h(F) | F \varepsilon F(\varepsilon X)\}$ is contained in some member of $\{h(F) | F \varepsilon F(\alpha X)\}$ and this occurs if and only if $\Gamma(\alpha X) \leq \Gamma(\varepsilon X)$. This says that Γ is a lattice isomorphism.

5.28 Remarks. The above proof demonstrates very effectively the advantages of using subfamilies of $C^*(X)$ to obtain compactifications. Many authors (including Magill for the above theorem) use quotient spaces of βX to produce a desired compactification. The disadvantage of this method is that proving the resultant quotient is Hausdorff is a major task. (Magill uses a full page of his article for this.) The only thing necessary to show when using the "evaluation compactification" techniques of 2.4 is that the subfamily of $C^*(X)$ which is being used separates points and closed sets. Usually this is completely routine. It is worth observing here that we needed X and Y locally compact in 5.27 in order to show this.

We will next be concerned with the following question. How far down in $K(X)$ can one expect a given $f \varepsilon C^*(X)$ to have an extension?

5.29 Theorem (Chandler and Gellar [1973]). Let $f \in C^*(X)$ have an extension f^α to $\alpha X \in K(X)$. There does not exist an element $\gamma X \in K(X)$ with $\gamma X < \alpha X$ to which f has an extension if and only if $f^\alpha|_{\alpha X \setminus X}$ is 1 - 1.

Proof. If f^α is not 1 - 1 on $\alpha X \setminus X$ we easily obtain $\gamma X < \alpha X$ by identifying some pair of points $x_1, x_2 \in \alpha X \setminus X$ for which $f^\alpha(x_1) = f^\alpha(x_2)$.

Suppose $\gamma X \leq \alpha X$ and f has extensions $f^\alpha : \alpha X \to \mathbb{R}$, $f^\gamma : \gamma X \to \mathbb{R}$ and suppose f^α is 1 - 1 on $\alpha X \setminus X$. There is a map $h: \alpha X \to \gamma X$ making

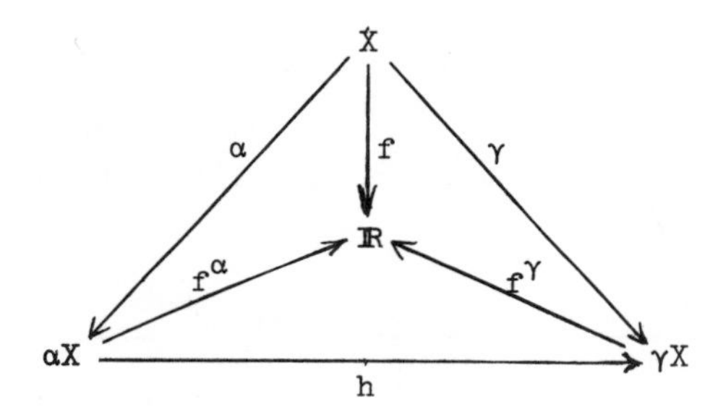

commute.

$f^\alpha = f^\alpha \circ h$ implies $h|_{\alpha X \setminus X}$ is 1 - 1. Now $h|_{X \subset \alpha X}$ is 1 - 1 and $h(\alpha X \setminus X) \subset \gamma X \setminus X$ by 1.30. Thus h is 1 - 1 from the compact space αX onto the Hausdorff space γX and is therefore a homeomorphism. We conclude that $\alpha X = \gamma X$.

5.30 Remarks. In view of this result we can see that if $\alpha X \setminus X$ is finite then there will always be functions in $C^*(X)$ which extend to αX and which extend to no smaller element of $K(X)$. It is somewhat surprising that this same result holds if $\alpha X \setminus X$ is countable. We need the fact that for any countable subset $Y_0 \subset Y$ there is a function in $C(Y)$ which is 1 - 1 on Y_0 (5.32). This was apparently first proved by Mrowka [1970] in a paper concerned with a separation property weaker than complete regularity, functional Hausdorffness. We give a different proof here, one which makes use of the fact that π is transcendental. This approach was used in Chandler and Gellar [1973].

5.31 Lemma. If $\{y_k\}_{k=1}^\infty$ is a sequence of distinct points in Y then for each positive integer n there is an $f_n \in C^*(Y)$ such that

(i) $f_n(Y) \subset [0,1]$

(ii) $f_n(y_p) = 0,\ 1 \le p < n$

(iii) $f_n(y_n) = 1$

(iv) $f_n(y_p)$ is rational if $p > n$.

Proof. By complete regularity there is a continuous $g_0\colon\ Y \to [0,1]$ such that $g_0(y_p) = 0,\ 1 \le p < n$, and $g_0(y_n) = 1$. We proceed inductively as follows. Suppose $g_0, g_1, \ldots, g_{k-1}$ have been defined and $g_{k-1}(y_{n+k}) = c$. If $c \in \{0, 1/2^{k-1}, 2/2^{k-1}, \ldots, (2^{k-1} - 1)/2^{k-1}, 1\}$, define $g_k = g_{k-1}$. Otherwise let $g_k = h_k \circ g_{k-1}$ where, if $c \in (p/2^{k-1}, (p + 1)/2^{k-1})$, then $h_k\colon\ [0,1] \to [0,1]$ is the homeomorphism which is the identity mapping on $[0, p/2^{k-1}] \cup [(p + 1)/2^{k-1}, 1]$, sends $[p/2^{k-1}, c]$ onto $[p/2^{k-1}, (2p + 1)/2^k]$, and sends $[c, (p + 1)/2^{k-1}]$ onto $[(2p + 1)/2^k, (p + 1)/2^{k-1}]$:

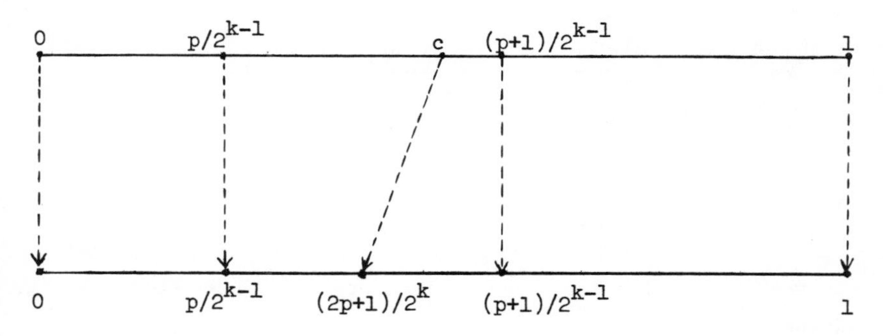

We claim that $\{g_k\}$ is a uniformly convergent sequence of functions from Y into $[0,1]$. This is an immediate consequence of the fact that $|g_k(x) - g_{k+p}(x)| \le |g_k(x) - g_{k+1}(x)| + |g_{k+1}(x) - g_{k+2}(x)| + \ldots + |g_{k+p-1}(x) - g_{k+p}(x)| \le 1/2^k + 1/2^{k+1} + \ldots + 1/2^{k+p-1} < 1/2^{k-1}$.

Let $f_n = \lim_{k\to\infty} g_k$. Then $f_n\colon\ Y \to [0,1]$ is continuous and $f_n(y_p) = 0$ for $1 \le p < n$ since $g_k(y_p) = 0$ for all k if $1 \le p < n$. $f_n(y_n) = 1$ since $g_k(y_n) = 1$ for all k. $f_n(y_p)$ is rational for $p > n$ since $g_k(y_p)$ is constant (and rational) for all $k \ge p - n$.

5.32 Theorem. If $\{y_n\}_{n=1}^{\infty}$ is a sequence of distinct points in Y then there is a continuous $f \in C^*(Y)$ such that $f(y_i) \ne f(y_j)$ if $i \ne j$.

Proof. Let $\{f_n\}_{n=1}^{\infty}$ be the sequence guaranteed by 5.31 and define

$f(y) = \sum_{n=1}^{\infty} (\pi/4)^n f_n(y)$. Clearly f is continuous. If $f(y_i) = f(y_j)$ for $i > j$ then $0 = \sum_{n=1}^{\infty} (\pi/4)^n f_n(y_i) - \sum_{n=1}^{\infty} (\pi/4)^n f_n(y_j) = \sum_{n=1}^{i} (\pi/4)^n [f_n(y_i) - f_n(y_j)]$ which, since $f_n(y_i) - f_n(y_j)$ is rational for all n, contradicts the transcendence of π.

5.33 Definition. For $f \in C^*(X)$ let $K(f)$ be the subset of $K(X)$ consisting of those compactifications to which f extends.

5.34 Corollary. If $|\beta X \setminus X| \leq \aleph_0$ then there exist f in $C^*(X)$ for which $K(f) = \{\beta X\}$.

Proof. By 5.32 there are functions in $C^*(X)$ for which $f^\beta|_{\beta X \setminus X}$ is 1 - 1. By 5.29, $K(f) = \{\beta X\}$ for these f.

5.35 Corollary. If $|\alpha X \setminus X| \leq \aleph_0$ then there exist f in $C^*(X)$ which have extensions to αX and to no smaller element of $K(X)$.

5.36 Remarks. Simple cardinality considerations dictate that if $|\alpha X \setminus X| > c$ then no function in $C^*(\alpha X)$ can be 1 - 1 on $\alpha X \setminus X$. In this case then $K(f)$ always contains elements smaller than αX (if it contains αX). This situation happens for βX if X is realcompact and not compact, for example, since then $|\beta X \setminus X| \geq 2^c$.

If $\aleph_0 < |\alpha X \setminus X| \leq c$ then the situation is too complicated to be resolved by cardinality arguments as the following two examples show.

5.37 Example. Let X be a space such that $\beta X \setminus X$ is homeomorphic to W (4.17). Then for no $f \in C^*(X)$ is it true that $K(f) = \{\beta X\}$ since no f^β can be 1 - 1 on $\beta X \setminus X$ (proof of 4.12). In this case $|\beta X \setminus X| = |W| = \aleph_1 \leq c$.

5.38 Example. Choose α so that $\aleph_\alpha > c$. Then if $X = [0,1] \times W(\omega_\alpha)$ it follows that $\beta X = [0,1] \times W(\omega_\alpha + 1)$ (4.15) and thus $\beta X \setminus X = [0,1] \times \{\omega_\alpha\}$. Define $f\colon X \to [0,1]$ to be the projection mapping onto the first coordinate. Then $f^\beta\colon \beta X \to [0,1]$ is the projection map also so that $f^\beta|_{\beta X \setminus X} = id_{[0,1]}$. In this case $|\beta X \setminus X| = c$.

5.39 Remarks. The following result, although more or less unrelated to compactifications, is an interesting and easy corollary to 5.32.

5.40 Theorem. If X is a countable, compact Hausdorff space then X is homeomorphic to a subset of $\mathbb{R}$.

Proof. By 5.32 there is a 1 - 1 map $f\colon X \to \mathbb{R}$. This is a homeomorphism by 1.11.

5.41 Major Problem #1. We have seen that $K(X)$ is a complete lattice if and only if X is locally compact (2.19). Additionally, $K(X)$ is a lattice if $\beta X \setminus X$ is discrete and C^*-embedded in βX (5.6). Finally, if X satisfies the first axiom of countability and is not locally compact, then $K(X)$ is not a lattice (5.12). Until recently, these were the significant results relating properties on X (or βX) with the question of whether or not $K(X)$ is a lattice. Fu-Chien Tzung, as part of his doctoral dissertation at N. C. State University (Tzung [1976]), proved that if $\beta X \setminus X$ is C^*-embedded in βX and if either

(i) $\alpha X \setminus X$ is realcompact for some $\alpha X \in K(X)$

or

(ii) $\alpha X \setminus X$ is a P-space for some $\alpha X \in K(X)$

then $K(X)$ is a lattice.

(A P-space is a space in which every point is a P-point (4.33) or equivalently, every G_δ-set is open. See Gillman and Jerison [1960].) He also constructed an example (using the technique of 4.18) which shows the condition that $\beta X \setminus X$ be C^*-embedded in βX is not sufficient by itself to guarantee that $K(X)$ is a lattice.

The recurrence of the condition that $\beta X \setminus X$ be C^*-embedded in βX is not accidental. It should be viewed as a natural generalization of local compactness: If X is locally compact then $\beta X \setminus X$ is compact and is therefore C^*-embedded in βX (Gillman and Jerison [1960], 3.11(e).) The other two conditions in Tzung's theorem are also generalizations of local compactness:

(i) If X is locally compact then $\alpha X \setminus X$ is compact (and hence realcompact) for all $\alpha X \in K(X)$.

(ii) If X is locally compact then $\omega X \setminus X$ is a P-space.

Tzung's results have one serious deficiency: they give no intrinsic conditions on X which imply that $K(X)$ is a lattice. A condition on X which guarantees that $\beta X \setminus X$ is realcompact has been given by Vaughan [1970] but apparently there are no conditions on X known to imply that $\beta X \setminus X$ is C^*-embedded in X (except local compactness). Neither do there seem to be results which would imply that $\alpha X \setminus X$ is a P-space for some $\alpha X \in K(X)$ (except local compactness).

To reiterate then, Major Problem #1 is to give intrinsic conditions on X which are necessary and/or sufficient for $K(X)$ to be a lattice.

§6: REMAINDERS - CARDINALITY CONSIDERATIONS

6.1 Remarks. In much of the previous material we have seen incidental results concerning remainders (i.e., $\alpha X \setminus X$ for $\alpha X \in K(X)$). To mention a few of these:

(i) $\alpha X \setminus X$ is compact if and only if X is locally compact.

(ii) For any X there is a Y with $\beta Y \setminus Y$ homeomorphic to X (4.17).

(iii) If X is compact and separable, there is an $\alpha N \in K(N)$ with $\alpha N \setminus N$ homeomorphic to X (4.20).

(iv) $|\beta N \setminus N| = 2^c$ (4.20).

(v) $\beta N \setminus N$ is not homeogeneous (4.36).

(vi) If D is an infinite discrete space then $\beta D \setminus D$ contains a subspace homeomorphic to βD (5.2).

(vii) If X and Y are locally compact then $\beta X \setminus X$ is homeomorphic to $\beta Y \setminus Y$ if and only if $K(X)$ is lattice isomorphic to $K(Y)$ (5.26, 27).

(viii) f^{α} is $1 - 1$ on $\alpha X \setminus X$ if and only if f extends to no compactification smaller than αX (5.29).

In this chapter we will attempt a more systematic study of problems concerning remainders, specifically cardinality considerations. Perhaps the first result in this direction (apart from (iv) above) is due to Hewitt [1948] who gave necessary and sufficient conditions for $|\beta X \setminus X| \leq 1$. This has been generalized by Firby [1971] whose results we will next consider. We will need the following definition.

6.2 Definition. $B_n(X) = \{f \in C^*(X) \mid$ there is a compact $K \subset X$ and $|f(X \setminus K)| \leq n\}$. $B_n(X)$ is uniformly dense in $C^*(X)$ if for $\varepsilon > 0$ and $f \in C^*(X)$ there is a $g \in B_n(X)$ with $|f(x) - g(x)| < \varepsilon$ for all $x \in X$. (This means $B_n(X)$ is dense in $C^*(X)$ where the topology on $C^*(X)$ is given by the sup metric.)

6.3 Definition. $A, B \subset X$ are completely separated if there is a function $f \in C^*(X)$ for which $f(A) = \{0\}$ and $f(B) = \{1\}$.

<u>6.4 Theorem</u>. The following are equivalent (n is finite).

(i) $|\beta X \setminus X| \leq n$.

(ii) For any $n + 1$ closed sets in X, completely separated in pairs, at least one is compact.

(iii) $B_n(X)$ is uniformly dense in $C^*(X)$.

<u>Proof</u>. <u>(i) implies (ii)</u>: Suppose $C_1,\ldots,C_{n+1}$ are closed sets in X, completely separated in pairs. Then $\overline{C}_1,\ldots,\overline{C}_{n+1}$ (closures in βX) are disjoint in pairs. As there are fewer than $n + 1$ points in $\beta X \setminus X$, at least one of these sets, say $\overline{C}_i$, is the same as C_i. Thus C_i is compact.

<u>(ii) implies (i)</u>: Suppose $x_1,\ldots,x_{n+1}$ are distinct points of $\beta X \setminus X$. Select compact neighborhoods $U_1,\ldots,U_{n+1}$ such that $x_i \in U_i$ and $U_i \cap U_j = \emptyset$ if $i \neq j$. Let $F_i = U_i \cap X$. It is a routine application of 5.8 to show that the sets $F_1,\ldots,F_{n+1}$ are completely separated in pairs. These sets are closed in X but none are compact since $x_i \in \overline{F}_i$ (closure in βX).

<u>(i) implies (iii)</u>: We may assume that $|\beta X \setminus X| = n$ since what we show is that $|\beta X \setminus X| = p$ implies that $B_p(X)$ is uniformly dense in $C^*(X)$. Thus, if $p \leq n$ then $B_n(X)$ is uniformly dense in $C^*(X)$ since $B_n(X) \supset B_p(X)$.

For $f' \in C^*(X)$ let $[m,M]$ be an interval containing $f'(X)$. If $\varepsilon' > 0$ let $\varepsilon = \varepsilon'/(M - m)$ and let $f = g \circ f'$ where $g: \mathbb{R} \to \mathbb{R}$ is defined by $g(t) = (t - m)/(M - m)$. Let $\overline{f}: \beta X \to [0,1]$ be the extension of f to βX and, if $\beta X \setminus X = \{a_1,\ldots,a_n\}$, let $\overline{f}(a_i) = b_i$. Relabel, if necessary, the b_i so that $\{b_1,\ldots,b_n\} = \{c_1,\ldots,c_p\}$ where $c_i \neq c_j$ if $i \neq j$. Let δ be the least element of the set

$$\{\varepsilon\} \cup \{|c_i - c_j| \mid i \neq j,\ 1 \leq i,\ j \leq p\}.$$

Set $U_i = \{x \in \beta X \mid |\overline{f}(x) - c_i| < \delta/2\}$ and select a closed neighborhood V_i (in βX) containing all a_j for which $\overline{f}(a_j) = c_i$ and which is contained in U_i. By 5.8 there is a $\overline{g}_i: \beta X \to [0,1]$ such that $\overline{g}_i(V_i) = \{0\}$, $\overline{g}_i(\beta X \setminus U_i) = \{1\}$. Let $g_i = \overline{g}_i|_X$ and define inductively

$$h_1 = \underline{c}_1 + (f - \underline{c}_1)g_1$$

and

$$h_i = \underline{c}_i + (h_{i-1} - \underline{c}_i)g_i,\ i = 2,\ldots,p.$$

(a) $h_i: X \to [0,1]$ **is continuous:**

Continuity is clear. Now $h_1 = \underline{c_1}[1 - g_1] + fg_1$ so that $h_1(x)$ is between c_1 and $f(x)$. Generally, $h_i = \underline{c_i}[1 - g_i] + h_{i-1}g_i$ so that $h_i(x)$ is between c_i and $h_{i-1}(x)$. Statement (a) then follows inductively.

(b) $h_i(x) = c_j$ whenever $x \in \text{int}(V_j) \cap X$ (interior in βX) and $1 \leq j \leq i$:

Let $W_j = X \cap \text{int}(V_j)$ (interior in βX). Since $g_1(x) = 0$ if $x \in V_1$, we have $h_1(x) = c_1$ if $x \in W_1$. Thus (b) holds for $i = 1$. Suppose (b) holds for $i = 1, 2, \ldots, k-1$. $h_k(x) = c_k + (h_{k-1}(x) - c_k)g_k(x)$. For $x \in W_j$, $1 \leq j \leq k - 1$, $g_k(x) = 1$ so that $h_k(x) = h_{k-1}(x) = c_j$ by the inductive hypothesis. If $x \in W_k$ then $g_k(x) = 0$ so that $h_k(x) = c_k$. Thus (b) holds for all i, j, $1 \leq i \leq p$, $1 \leq j \leq i$.

(c) $h_i(x) = f(x)$ if $x \in X \setminus \bigcup_{j=1}^{i} U_j$:

If $x \notin U_j$, $1 \leq j \leq i$, then $x \notin V_j$, $1 \leq j \leq i$, so that $g_j(x) = 1$. Thus, by definition $h_i(x) = h_{i-1}(x) = \ldots = h_1(x) = f(x)$.

(d) $|h_p(x) - f(x)| < \varepsilon$ for all $x \in X$:

From (b) and (c) all we need show is that $|h_p(x) - f(x)| < \varepsilon$ for $x \in T_i = X \cap (U_i \setminus V_i)$, $i = 1, 2, \ldots, p$. Now for $x \in T_1$ we have

$$\begin{aligned} |h_1(x) - f(x)| &= |c_1 + (f(x) - c_1)g_1(x) - f(x)| \\ &= |c_1 - f(x)|\ |1 - g_1(x)| \\ &\leq \delta/2 \cdot 1 < \varepsilon. \end{aligned}$$

Inductively, suppose that $|h_i(x) - f(x)| < \varepsilon$ whenever $x \in T_j$, $1 \leq j \leq i$, for $i = 1, 2, \ldots, k-1$. For $x \in T_j$, $1 \leq j \leq k - 1$ we have

$$h_k(x) = c_k + (h_{k-1}(x) - c_k)g_k(x) = h_{k-1}(x)$$

since $g_k(x) = 1$. Thus, by the inductive hypothesis $|h_k(x) - f(x)| < \varepsilon$. If $x \in T_k$, then by (c) $h_{k-1}(x) = f(x)$ so that

$$\begin{aligned} |h_k(x) - f(x)| &= |c_k + (f(x) - c_k)g_k(x) - f(x)| \\ &= |c_k - f(x)|\ |1 - g_k(x)| \\ &\leq \delta/2 \cdot 1 < \varepsilon. \end{aligned}$$

We have shown then, for $k = p$ that $|h_p(x) - f(x)| < \varepsilon$ for $x \in T_j$, $1 \leq j \leq p$.

(e) $h_p \in B_p(X)$:

The set $Y = \beta X \setminus \bigcup_{i=1}^{p} \text{int}(V_i)$ is compact and since $\beta X \setminus X = \{a_1, \ldots, a_n\} \subset \bigcup_{i=1}^{p} \text{int}(V_i)$ we have $Y \subset X$. On $X \setminus Y$ h_p takes on p values, $c_1, \ldots, c_p$. Therefore, $h_p \in B_p(X)$.

(f) $B_n(X)$ is uniformly dense in $C^*(X)$:

For our original f' and $\varepsilon' > 0$ observe that $|h_p'(x) - f'(x)| < \varepsilon'$ for all $x \in X$, if $h_p' = k \circ h_p$ where $k\colon \mathbb{R} \to \mathbb{R}$ is defined by $k(t) = (M - m)t + m$. $h_p' \in B_p(X) \subset B_n(X)$.

<u>(iii) implies (i)</u>: Suppose $\beta X \smallsetminus X \supset \{a_1, \ldots, a_{n+1}\}$. Choose pairwise disjoint open (in βX) neighborhoods U_i and closed neighborhoods V_i with $a_i \in V_i \subset U_i$. By 5.8 there is a function $f_i \in C^*(\beta X)$ with $f_i(V_i) = i$ and $f_i(\beta X \smallsetminus U_i) = 0$, $f_i(\beta X) \subset [0,i]$. Let $f = \sum_{i=1}^{n+1} f_i$. Then $f(a_i) = i$. For any compact set $K \subset X$ there are open sets $O_1, \ldots, O_{n+1}$ in X such that $O_i \cap O_j = \emptyset$ if $i \neq j$ and $O_i \cap K = \emptyset$. **Thus,** it is clear that no function taking only n values on $X \smallsetminus K$ can be within ε of f for suitably chosen $\varepsilon > 0$. We conclude that $B_n(X)$ is not uniformly dense in $C^*(X)$.

<u>6.5 Corollary</u>. The following are equivalent for finite n.

(i) $|\beta X \smallsetminus X| = n$

(ii) There are n closed, non-compact sets in X completely separated in pairs but no such collection of $n + 1$ sets exists.

(iii) $B_n(X)$ is uniformly dense in $C^*(X)$ and $B_{n-1}(X)$ is not uniformly dense in $C^*(X)$.

<u>6.6 Remarks</u>. If $\beta X \smallsetminus X$ is finite then $K(X)$ is also finite and $|K(X)|$ is precisely the same as the number of distinct equivalence relations on $\beta X \smallsetminus X$. (This is immediate from the two facts (i) $\beta X \smallsetminus X$ is discrete and so an equivalence relation determines a Hausdorff quotient space and conversely and (ii) each member of $K(X)$ is a quotient space of βX determined by an equivalence relation on $\beta X \smallsetminus X$ and conversely by 2.16.) The number of distinct equivalence relations on a finite set with n elements is a well-studied combinatorial object, called by Riordan the Bell number

$$B(n) = \sum_{r=1}^{n} \sum_{k=0}^{r} \frac{(-1)^k}{r!} \binom{r}{k} (r - k)^n.$$

A table of values of $B(n)$, $1 \leq n \leq 74$, is given in Levine and Dalton [1962]. They also give reference to other occurences of it in the combinatorial literature. For $1 \leq n \leq 6$ we have the following values for $B(n)$: 1, 2, 5, 15, 52, 203. For $n = 74$ (the largest value in their table) $B(n) > 5 \times 10^{78}$.

Notice that these results are in direct contrast to the situation if $\beta X \setminus X$ is not finite: given any infinite cardinal number n there is an X for which $|K(X)| = n$ (5.4). For finite n only those which are Bell numbers can be the cardinal number of some $K(X)$.

The prototype X_n for which $|\beta X_n \setminus X_n| = n$ might well be $W \times \{1, 2, \ldots, n\}$. If we represent W as a right-open interval, ———o, then $X_n = W \times \{1, 2, \ldots, n\}$ could be represented

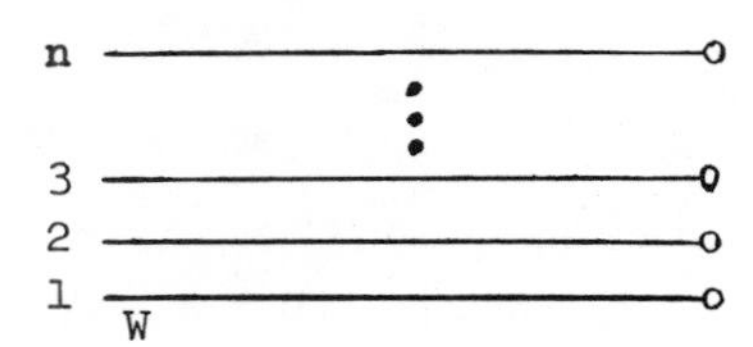

We could thus represent βX_n by using right-closed intervals:

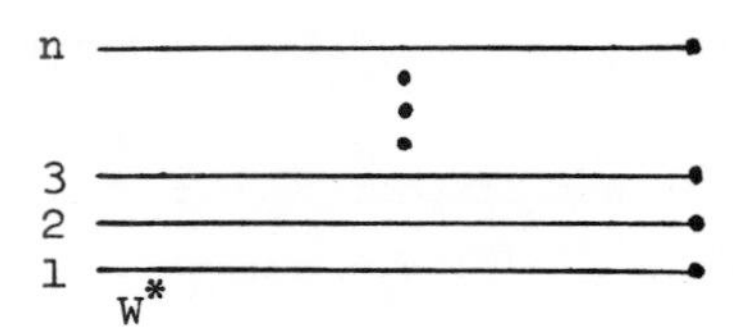

For small n we may even represent $K(X_n)$:

n = 1: βX_1 ———•

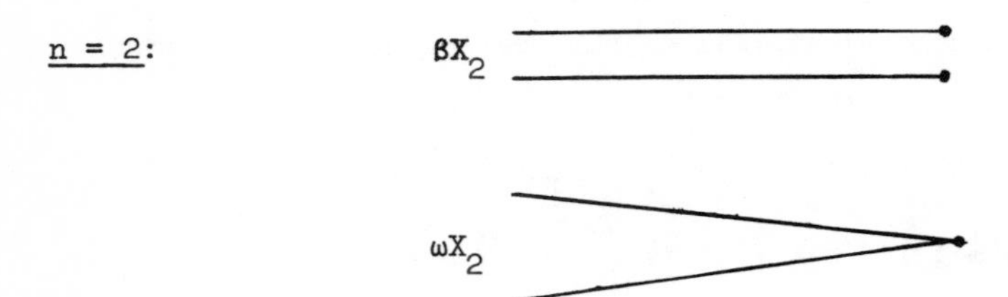

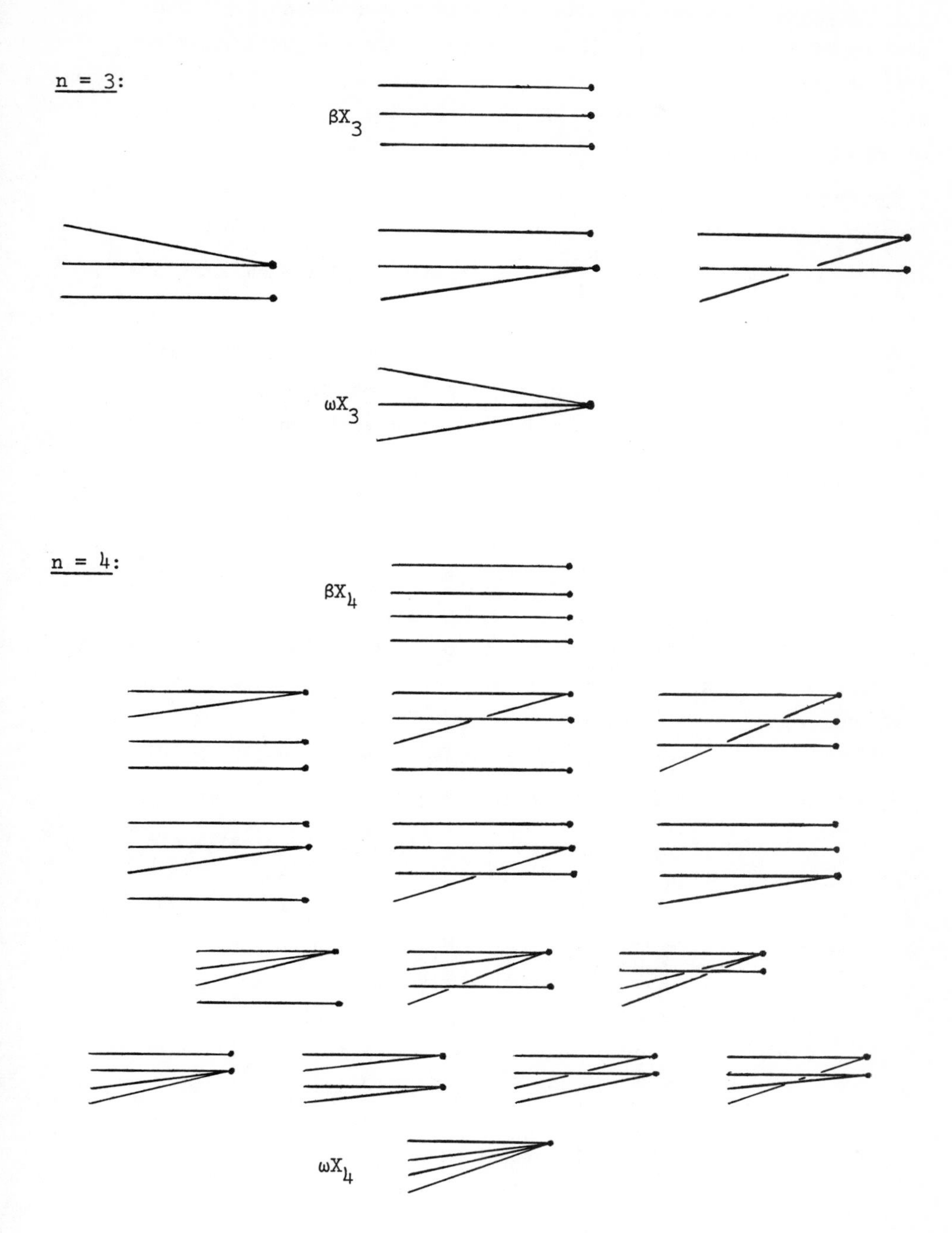
n = 3:
βX_3
ωX_3
n = 4:
βX_4
ωX_4

<u>6.7 Remarks</u>. Thus far, we have been primarily concerned with the problem of when $\beta X \setminus X$ is finite. We next consider the problem of determining if <u>some</u> remainder of X is finite; i.e., does there exist an element $\alpha X \in K(X)$ for which $\alpha X \setminus X$ is finite? Magill [1965] gives necessary and sufficient conditions for this to occur.

<u>6.8 Theorem</u>. For some $\alpha X \in K(X)$, $|\alpha X \setminus X| = n$ if and only if X is locally compact and contains n open, non-empty pairwise disjoint subsets $\{G_i\}_{i=1}^n$ such that $K = X \setminus \bigcup_{i=1}^n G_i$ is compact but for each i, $K \cup G_i$ is not compact.

<u>Proof</u>. If $\alpha X \setminus X = \{x_1, \ldots, x_n\}$ select open sets in αX $G_1', \ldots, G_n'$ such that $x_i \in G_i'$ and $G_i' \cap G_j' = \emptyset$ if $i \neq j$. Let $G_i = G_i' \cap X$. Then $\{G_i\}_{i=1}^n$ is a family of pairwise disjoint, non-empty open sets and $K = X \setminus \bigcup_{i=1}^n G_i = \alpha X \setminus \bigcup_{i=1}^n G_i'$ is compact. Now

$$K \cup G_i = X \setminus [G_1 \cup \ldots \cup G_{i-1} \cup G_{i+1} \cup \ldots \cup G_n]$$
$$= \{\alpha X \setminus [G_1' \cup \ldots \cup G_{i-1}' \cup G_{i+1}' \cup \ldots \cup G_n']\} \cap \{\alpha X \setminus \{x_i\}\}.$$

If $K \cup G_i$ were compact then it would be a closed subset of αX. Then $G = G_1' \cup \ldots \cup G_{i-1}' \cup G_{i+1}' \cup \ldots \cup G_n' \cup \{x_i\}$ would be an open set in αX so that $\{x_i\} = G_i' \cap G$ is an isolated point in αX, a contradiction.

Conversely, suppose $\{G_i\}_{i=1}^n$ is a family of non-empty, pairwise disjoint open subsets of X for which $K = X \setminus \bigcup_{i=1}^n G_i$ is compact and $K \cup G_i$ is not compact for each i. Let $\{p_i\}_{i=1}^n$ be a collection of n distinct points not in X. $\mathcal{H}_i = \{O \mid O$ is open in X and $[K \cup G_i] \cap [X \setminus O]$ is compact$\}$ is closed under finite intersections so that the family $\mathcal{H}_i^* = \{O \cup \{p_i\} \mid O \in \mathcal{H}_i\}$ forms a neighborhood base at p_i in $\alpha X = X \cup \{p_1, \ldots, p_n\}$. We then give αX the topology with base $\mathcal{T} \cup \mathcal{H}_1^* \cup \ldots \cup \mathcal{H}_n^*$ where $\mathcal{T}$ is the topology on X.

It is routine to verify that αX with this topology is Hausdorff. In particular, $p_i \in G_i \cup \{p_i\} \in \mathcal{H}_i^*$, $p_j \in G_j \cup \{p_j\} \in \mathcal{H}_j^*$ and $[G_i \cup \{p_i\}] \cap [G_j \cup \{p_j\}] = \emptyset$. Likewise, separating $x \in X$ from p_i follows immediately from the hypothesis that X is locally compact.

To see that αX is compact simply observe that for any open covering, once the points $p_1, \ldots, p_n$ are covered, all that remains is a compact set (by the definition of the neighborhood base at p_i).

For any open set $O_i = O \cup \{p_i\}$ containing p_i we have that $[K \cup G_i] \cap [X \setminus O]$ is compact whereas $[K \cup G_i] \cap X$ is not compact. Thus $X \cap O \neq \emptyset$ and we conclude that X is dense in αX.

<u>6.9 Remarks</u>. It is tempting to attempt to obtain αX above by using the set $F = B_n(X) \subset C^*(X)$ (6.2) to construct the compactification $e_F X$ (2.4). With the stated hypothesis, it is easily verified that F separates points and closed sets and so determines a compactification $e_F X$. What is not apparent is that $|e_F X \setminus X| = n$. To show this we would need to know that $B_n(X) \setminus B_{n-1}(X) \neq \emptyset$. This seems difficult.

This suggests the following (seemingly unsolved) question. Call two subsets $F, G \subset C^*(X)$ <u>equivalent</u> if $e_F X \approx e_G X$. What are necessary and sufficient conditions for F and G to be equivalent?

<u>6.10 Corollary</u>. $\mathbb{R}$ has no 3-point compactification; i.e., for no $\alpha\mathbb{R}$ is it true that $|\alpha\mathbb{R} \setminus \mathbb{R}| = 3$.

<u>Proof</u>. If $\alpha\mathbb{R}$ exists for which $|\alpha\mathbb{R} \setminus \mathbb{R}| = 3$ then by 6.8 $\mathbb{R} = K \cup G_1 \cup G_2 \cup G_3$ where K is compact, G_i is open, $K \cup G_i$ is not compact, $K \cap G_i = \emptyset$, and $G_i \cap G_j = \emptyset$ for $i, j = 1, 2, 3, i \neq j$.

Since $K \cup G_i$ is closed and not compact it is not bounded and hence G_i is not bounded for $i = 1, 2, 3$. Let the first and last points of K be denoted m and M, respectively. Then either two of the G_i have points greater than M or two of the G_i have points less than m. For definiteness, assume there exists $g_1 \in G_1$ and $g_2 \in G_2$ with $M < g_1 < g_2$. Let p be the least upper bound of the set $\{x \in G_1 | x < g_2\}$. Then $p \notin K$ since $p > M$. Also $p \notin G_i$ for $i = 1, 2, 3$ since the G_i are open and hence if one contains p it contains an interval containing p. This would contradict the definition of p. This contradiction assures us that the initial assumption of the existence of an $\alpha\mathbb{R}$ with $|\alpha\mathbb{R} \setminus \mathbb{R}| = 3$ is false.

<u>6.11 Corollary</u>. If $n > 1$ there exists no $\alpha\mathbb{R}^n$ with $|\alpha\mathbb{R}^n \setminus \mathbb{R}^n| = 2$.

<u>Proof</u>. If so, then $\mathbb{R}^n = K \cup G_1 \cup G_2$ with K compact, $K \cup G_1$ and $K \cup G_2$ not compact, $K \cap G_1 = \emptyset$, $K \cap G_2 = \emptyset$, and $G_1 \cap G_2 = \emptyset$.

As in 6.10 K is bounded and G_1 and G_2 are not. Let B be a ball in $\mathbb{R}^n$ with $K \subset B$. There exist $g_1 \in G_1 \setminus B$, $g_2 \in G_2 \setminus B$. Let $f\colon [0,1] \to \mathbb{R}^n \setminus B$ be a homeomorphism with $f(0) = g_1$, $f(1) = g_2$. Let $A = \{t \in [0,1] | f([0,t]) \subset G_1\}$ and suppose $p = \sup A$. As in 6.10, $f(p) \notin G_1$, $f(p) \notin G_2$ since they are open and $f(p) \notin K$ since

$f(p) \in \mathbb{R}^n \setminus B$ and $K \subset B$.

6.12 Lemma. If X has an n-point compactification and does not have an $(n + 1)$-point compactification, then all n-point compactifications of X are equivalent.

Proof. Suppose $|\alpha X \setminus X| = n = |\gamma X \setminus X|$. Then $|F(\alpha X)| = n = |F(\gamma X)|$ (see 5.15). If $\alpha X \not\approx \gamma X$ then $F(\alpha X) \neq F(\gamma X)$ by 5.17. Suppose $F(\alpha X) = \{A_1, \ldots, A_n\}$ and $F(\gamma X) = \{G_1, \ldots, G_n\}$. Then some element of $F(\alpha X)$, say A_i is equal to no G_j, $j = 1, \ldots, n$. Since $F(\alpha X)$ and $F(\gamma X)$ are partitions of $\beta X \setminus X$ we must have A_i intersecting at least two of the elements of $F(\gamma X)$, say $A_i \cap G_k \neq \emptyset \neq A_i \cap G_\ell$, or, in the case A_i is properly contained in some G_j, we must have G_j intersecting at least two of the elements of $F(\alpha X)$, say $G_j \cap A_i \neq \emptyset \neq G_j \cap A_p$.

In the first case, create a partition of $\beta X \setminus X$ as follows: $\{A_1, A_2, \ldots, A_{i-1}, A_i \cap G_k, A_i \cap (\bigcup_{j \neq k} G_j), A_{i+1}, \ldots, A_n\}$. In the second case, our partition of $\beta X \setminus X$ will be $\{G_1, \ldots, G_{j-1}, G_j \cap A_i, G_j \cap (\bigcup_{k \neq i} A_k), G_{j+1}, \ldots, G_n\}$.

Thus, in either case, we have a partition of $\beta X \setminus X$ having $n + 1$ elements. If δX is the corresponding compactification then $|\delta X \setminus X| = n + 1$.

6.13 Lemma. If X has an n-point compactification and $1 \leq p \leq n$, then X has a p-point compactification.

Proof. If $|\alpha X \setminus X| = n$ and $F(\alpha X) = \{A_1, \ldots, A_n\}$, form the partition of $\beta X \setminus X$: $\{A_1, \ldots, A_{p-1}, \bigcup_{i=p}^{n} A_i\}$. This has p elements so that the corresponding compactification δX is such that $|\delta X \setminus X| = p$.

6.14 Theorem. (a) The only compactifications of $\mathbb{R}$ with finite remainder are $\omega\mathbb{R}$ and $\tau\mathbb{R} = [-\infty, \infty]$.

(b) If $n > 1$, the only compactification of $\mathbb{R}^n$ with finite remainder is $\omega\mathbb{R}^n$.

Proof. These two results are immediate from 6.10 and 6.11 in view of 6.12 and 6.13.

6.15 Remarks. We end our consideration of compactifications with finite remainders with a result which leads naturally to the next thing we take up, compactifications with countably infinite remainders. We need a lemma due to Thrivikraman [1972] for this.

<u>6.16 Lemma</u>. Let X be locally compact. The following are equivalent:

(a) X has a 2-point compactification, τX.

(b) X has a compactification αX with $\alpha X \setminus X$ disconnected.

(c) $\beta X \setminus X$ is disconnected.

<u>Proof</u>. (a) implies (b) trivially. (b) implies (c) since if $\alpha X \setminus X = A_1 \cup A_2$ with $A_1 \cap \overline{A}_2 = \emptyset = \overline{A}_1 \cap A_2$ and π_α is the quotient map, π_α: $\beta X \to \alpha X$, then $\pi_\alpha^{-1}(A_1) = B_1$, $\pi_\alpha^{-1}(A_2) = B_2$ is a separation of $\beta X \setminus X$.

To see that (c) implies (a) suppose $\beta X \setminus X = B_1 \cup B_2$ with $B_1 \cap \overline{B}_2 = \emptyset = \overline{B}_1 \cap B_2$. Then $\{B_1, B_2\}$ is a partition of $\beta X \setminus X$ and the associated compactification has a remainder with only two points.

<u>6.17 Theorem</u>. If X is locally compact and there is an $\alpha X \in K(X)$ with $1 < |\alpha X \setminus X| \leq \aleph_0$ then X has a 2-point compactification.

<u>Proof</u>. $\alpha X \setminus X$ is compact and so by 5.40 **is** homeomorphic to a subset of $\mathbb{R}$. Thus, $\alpha X \setminus X$ is not connected (no countable, nondegenerate subset of $\mathbb{R}$ is connected). By 6.16, X has a 2-point compactification.

<u>6.18 Corollary</u>. For $n > 1$, the only compactification of $\mathbb{R}^n$ with countable remainder is $\omega\mathbb{R}^n$.

<u>Proof</u>. Immediate from 6.17 and 6.14 (b).

<u>6.19 Remarks</u>. We will next be concerned with compactifications having a countable remainder. One of the results we will need (6.24 below) is of interest in itself here since it concerns the cardinality of a remainder. We will need a sequence of simple lemmas first. Recall that a subspace A of a space X is <u>C-embedded</u> if each continuous f: $A \to \mathbb{R}$ has an extension f^*: $X \to \mathbb{R}$.

<u>6.20 Lemma</u>. A closed subset of $\mathbb{R}$ is C-embedded.

<u>Proof</u>. Let F be closed in $\mathbb{R}$ and f: $F \to \mathbb{R}$. Write $\mathbb{R} \setminus F$ as the union of maximal, disjoint open intervals: $\mathbb{R} \setminus F = \bigcup_{\alpha \in A} I_\alpha$, where $I_\alpha = (a_\alpha, b_\alpha)$. Define f^*: $\mathbb{R} \to \mathbb{R}$ by

$$f^*(x) = f(a_\alpha) + (x - a_\alpha)(f(a_\alpha) - f(b_\alpha)) / (a_\alpha - b_\alpha)$$

for each $x \in I_\alpha$ and for each $\alpha \in A$. (If $a_\alpha = -\infty$ for some α, then define $f^*(x) = f(b_\alpha)$ for all $x \in I_\alpha$. Similarly, if $b_\alpha = +\infty$ for some α, define $f^*(x) = f(a_\alpha)$ for all $x \in I_\alpha$.)

6.21 Lemma. Let $F \subset X$ be closed and suppose that for some $f \in C(X)$, $f|_F$ is a homeomorphism of F onto a closed subset of $\mathbb{R}$. Then F is C-embedded.

Proof. For arbitrary $g\colon F \to \mathbb{R}$ we have $g \circ f|_F^{-1}\colon f(F) \to \mathbb{R}$ where $f(F)$ is closed in $\mathbb{R}$. By 6.20 this function has an extension $h\colon \mathbb{R} \to \mathbb{R}$. Define $g^*\colon X \to \mathbb{R}$ by $g^* = h \circ f$. For $x \in F$ we have $g^*(x) = (h \circ f)(x) = (g \circ f|_F^{-1} \circ f)(x) = g(x)$, since for $x \in F$ we have $(f|_F^{-1} \circ f)(x) = x$. Thus g^* extends g.

6.22 Lemma. X is pseudocompact (every continuous $f\colon X \to \mathbb{R}$ is bounded) if and only if X contains no C-embedded copy of N.

Proof. Suppose X contains a C-embedded copy of N, say $S = \{x_1, x_2, \ldots\}$. Define $f\colon S \to \mathbb{R}$ by $f(x_i) = i$. Then the extension of f, $f^*\colon X \to \mathbb{R}$ is not bounded; i.e., X is not pseudocompact.

Conversely, if X is not pseudocompact obtain a subset $S = \{x_1, x_2, \ldots\}$ by first choosing x_1 so that $f(x_1) \geq 1$ (where we assume that $f\colon X \to \mathbb{R}$ is unbounded above. If the unbounded functions from $X \to \mathbb{R}$ are unbounded below, modify the following argument in the obvious manner). For each $i > 1$ choose x_i so that $f(x_i) \geq f(x_{i-1}) + 1$. Then S is a copy of N and $f|_S$ is a homeomorphism onto $f(S)$, a closed subset of $\mathbb{R}$. By 6.21 S is C-embedded.

6.23 Lemma. If S is a C-embedded copy of N in X then S is closed in X.

Proof. Let x be an arbitrary point of $X \setminus S$ and if $S = \{x_1, x_2, \ldots\}$, define $f\colon S \to \mathbb{R}$ by $f(x_i) = i$. If $f^*\colon X \to \mathbb{R}$ is the extension of f to X, choose a neighborhood U of $f^*(x)$ in $\mathbb{R}$ so that U contains no more than one integer. Then $f^{*-1}(U)$, a neighborhood of x in X, can contain no more than one point of S. Thus, x is not a limit point of S.

6.24 Theorem. If $|\beta X \setminus X| < 2^c$ then X is pseudocompact.

Proof. If X is not pseudocompact then by 6.22 X has a C-embedded copy of N as a subset S. Then $\beta S = \overline{S}^{\beta X}$ (closure in βX). Now

$$\overline{S}^{\beta X} \cap X = \overline{S}^X = S \quad \text{(by 6.23)}$$

so that $\beta S \setminus S \subset \beta X \setminus X$. Then

$$|\beta X \setminus X| \geq |\beta S \setminus S| = |\beta N \setminus N| = 2^c \quad (4.20).$$

<u>6.25 Lemma</u>. If (X,d) is a compact metric space and $Y \subset X$ is pseudocompact then Y is closed (hence compact).

<u>Proof</u>. If Y is not closed then some sequence $\{y_n\} \subset Y$ converges to $x \in X \smallsetminus Y$. Define $f\colon Y \to \mathbb{R}$ by $f(y) = 1/d(x,y)$. Clearly f is unbounded and continuous.

<u>6.26 Theorem</u>. (Okuyama [1971]). The following are equivalent.

(a) $|\beta X \smallsetminus X| \leq \aleph_0$

(b) X is pseudocompact and there is a continuous map $f\colon X \to [0,1]$ such that the set $Y_o = \{y \in f(X) | f^{-1}(y)$ is not compact$\}$ is countable and $|\mathrm{cl}_{\beta X} f^{-1}(y) \smallsetminus f^{-1}(y)| = 1$ if $y \in Y_o$.

<u>Proof</u>. (a) implies (b): Let $\beta X \smallsetminus X = \{p_1, p_2, \ldots\}$ where $p_i \neq p_j$ if $i \neq j$. By 6.24 X is pseudocompact. By 5.32 there is a map $g\colon \beta X \to [0,1]$ such that $g(p_i) \neq g(p_j)$ if $i \neq j$. Let $f = g|_X$. If $Y = f(X)$ then Y is a pseudocompact subset of $[0,1]$ so that by 6.25 Y is compact. Now X is dense in βX so that $g(X)$ is dense in Y. Therefore, $g(\beta X) = Y$. Let $q_i = g(p_i)$. For $y \in Y$ we have $g^{-1}(y) \cap X = f^{-1}(y)$ and $g^{-1}(y)$ is a closed subset of βX. Thus,

$$\mathrm{cl}_{\beta X} f^{-1}(y) \subset g^{-1}(y).$$

If $y \neq q_i$ for any i, then $g^{-1}(y) \subset X$ so that $f^{-1}(y)$ is compact. Thus

$$Y_o = \{y \in Y | f^{-1}(y) \text{ is not compact}\} \subset \{q_1, q_2, \ldots\}$$

and so Y_o is countable.

If $y = q_i$ for some i and $f^{-1}(y)$ is not compact then $\mathrm{cl}_{\beta X} f^{-1}(y) \neq f^{-1}(y)$ so that $\mathrm{cl}_{\beta X} f^{-1}(y)$ contains at least one point of $g^{-1}(y) \cap (\beta X \smallsetminus X)$. However $g^{-1}(y) = f^{-1}(y) \cup \{p_i\}$ (remember $y = q_i$). Thus $\mathrm{cl}_{\beta X} f^{-1}(y) = f^{-1}(y) \cup \{p_i\}$ so that

$$|\mathrm{cl}_{\beta X} f^{-1}(y) \smallsetminus f^{-1}(y)| = 1 \text{ if } y \in Y_o.$$

(b) implies (a): Let $f\colon X \to [0,1]$ and let $Y_o = \{y \in Y | f^{-1}(y)$ is not compact$\}$ be countable and for each $y \in Y_o$ suppose $|\mathrm{cl}_{\beta X} f^{-1}(y) \smallsetminus f^{-1}(y)| = 1$. Let $\mathrm{cl}_{\beta X} f^{-1}(y_i) = f^{-1}(y_i) \cup \{p_i\}$ for each $y_i \in Y_o = \{y_1, y_2, \ldots\}$. As above, if $f^{-1}(y)$ is compact, then $f^{\beta-1}(y) = f^{-1}(y)$ where $f^{\beta}\colon \beta X \to f(X)$ is the extension of f ($f(X)$ is compact). Now if $Y = f(X)$, $\beta X = f^{\beta-1}(Y) = \bigcup_{y \in Y} f^{\beta-1}(y) = \bigcup_{y \in Y} \mathrm{cl}_{\beta X} f^{-1}(y) = [\bigcup_{y \in Y} f^{-1}(y)] \cup \{p_1, p_2, \ldots\}$.

Thus $\beta X \setminus X = \{p_1, p_2, \ldots\}$ and so is countable.

6.27 Remarks. The preceeding is actually a slight improvement of Okuyama's result: In his theorem Y is a subset of the Hilbert cube. This seems to be the best result available giving necessary and sufficient conditions for $\beta X \setminus X$ to be countable. No results are available giving necessary and sufficient conditions for $|\beta X \setminus X| = c$ or $\aleph_1$ (or any cardinal between $\aleph_0$ and 2^c.) In view of 6.24 such an X would have to be pseudocompact. More needs to be done. The key ingredient in attempting to obtain a result analogous to 6.26 would be to find a space Z for which pseudocompact subsets are compact, and for which a map $g: \beta X \to Z$ could be defined which is $1 - 1$ on $\beta X \setminus X$. Of course, this may well be impossible or, if possible, impractical. In fact, 6.26 as it is seems to strain the limits of practicality.

We next look at a result of Magill [1966] giving necessary and sufficient conditions for a space to have some compactification with a countable remainder. First we need some standard definitions and lemmas concerning totally disconnected spaces. These can be found in Gillman and Jerison [1960] for example.

6.28 Definitions. For $S \subset X$ a partition of S is a finite collection of disjoint relatively open and closed subsets of S whose union is S. A set is connected if it has no partition with more than one element. A space X is totally disconnected if no connected subset of X contains more than one point. The component containing a given point $x \in X$ is the union of all connected subsets of X which contain x. Components are always connected and closed.

6.29 Lemma. Let $\mathcal{K}$ be a family of compact subsets of X and let $\{K_1, K_2\}$ be a partition of $\bigcap_{K \in \mathcal{K}} K$. Then there is a finite subcollection $\mathcal{H} \subset \mathcal{K}$ and a partition $\{H_1, H_2\}$ of $\bigcap_{H \in \mathcal{H}} H$ such that $H_i \supset K_i$, $i = 1,2$.

Proof. K_1 and K_2 are disjoint and closed in $\bigcap_{K \in \mathcal{K}} K$, a compact set. Thus there are disjoint open sets, U_1, U_2 with $K_i \subset U_i$. Now $\{U_1 \cup U_2\} \cup \{X \setminus K\}_{K \in \mathcal{K}}$ is an open cover of X and hence of $K_1' \in \mathcal{K}$. Thus, there is a finite family $\mathcal{H}' \subset \mathcal{K}$ such that $K_1' \subset U_1 \cup U_2 \cup X \setminus K_2' \cup \ldots \cup X \setminus K_n'$ where $\mathcal{H}' = \{K_2', \ldots, K_n'\}$. It follows that $K_1' \cap K_2' \cap \ldots \cap K_n' \subset U_1 \cup U_2$. Let $\mathcal{H} = \{K_1', \ldots, K_n'\}$ and let $H_i = K_1' \cap \ldots \cap K_n' \cap U_i$, $i = 1,2$.

6.30 Lemma. The component of a point x in a compact space X is the intersection of all open and closed sets in X which contain x.

Proof. Let $\mathcal{K}$ be the family of open and closed sets containing x. If $\{K_1, K_2\}$ is a partition of $\bigcap_{K\in\mathcal{K}} K$ with $x \in K_1$ then let H_1 and H_2 be as in 6.29. $H_1 \cup H_2 = \bigcap_{H\in\mathcal{H}} H$, an intersection of finitely many open and closed sets, is open and closed. Thus H_1 and H_2 are open and closed in X. Since $x \in K_1 \subset H_1$ we have $\bigcap_{K\in\mathcal{K}} K \subset H_1$ (since $K_1 \in \mathcal{K}$). Thus $\{K_1, K_2\}$ is not a partition of $\bigcap_{K\in\mathcal{K}} K$. We conclude that $\bigcap_{K\in\mathcal{K}} K$ is connected.

If A is any set containing $\bigcap_{K\in\mathcal{K}} K$ and $y \in A \setminus \bigcap_{K\in\mathcal{K}} K$ then there is an open and closed set containing $\bigcap_{K\in\mathcal{K}} K$ and not containing y. Thus A is not connected. We conclude that $\bigcap_{K\in\mathcal{K}} K$ is the component of X which contains x.

6.31 Lemma. A compact space X is totally disconnected if and only if X has a base of open and closed sets.

Proof. If X is totally disconnected suppose $x \in X$ and U is a neighborhood of x. The family $\mathcal{K}$ of all open and closed sets in X containing x is a collection of compact sets whose intersection is contained in U by 6.30. As we saw in the proof of 6.29 there is a finite collection $\{K_1,\ldots,K_n\} \subset \mathcal{K}$ for which $K = K_1 \cap\ldots\cap K_n \subset U$. Since this is a finite intersection of open and closed sets, K is an open and closed set in X containing x and contained in U.

Conversely, if X has a base of open and closed sets then if x and y are distinct points of X they are contained in disjoint open and closed sets. Thus no component in X contains both x and y.

6.32 Theorem. The following are equivalent.

(a) X is locally compact and $\beta X \setminus X$ has infinitely many components.

(b) X is locally compact and there is an $\alpha X \in K(X)$ such that $\alpha X \setminus X$ is infinite and totally disconnected.

(c) X is locally compact and $|\gamma X \setminus X| = \aleph_0$ for some $\gamma X \in K(X)$.

(d) X has an n-point compactification for each n.

Proof. (a) implies (b): Let $F = \{f \in C^*(X) | f^\beta$ is constant on each component of $\beta X \setminus X\}$. Since $\beta X \setminus X$ is compact, it is routine to verify that F separates points and closed sets in X. (Actually, $F_0 \subset F$ separates points and closed sets in X where F_0 is the set of elements of $C^*(X)$ whose extension to βX is constant on $\beta X \setminus X$.) Let

$\alpha X = e_F X$ (2.4). Let $\pi_\alpha : \beta X \to \alpha X$ be the quotient map.

If A is a component of $\beta X \setminus X$ then $\pi_\alpha(A)$ is a point of $\alpha X \setminus X$. This holds since if $p, q \in A$ then **for all** $f \in F$ **we have**

$$f^\alpha \circ \pi_\alpha(p) = f^\beta(p) = f^\beta(q) = f^\alpha \circ \pi_\alpha(q).$$

It follows from the definition of $\alpha X = e_F X$ that $\pi_\alpha(p) = \pi_\alpha(q)$. Since p and q were arbitrary points of A, our assertion that $\pi_\alpha(A)$ is a point holds.

We next claim that **if** A_1 **and** A_2 are distinct components of $\beta X \setminus X$ then $\pi_\alpha(A_1) \neq \pi_\alpha(A_2)$. This follows from the fact that in F there is a function f such that $f^\beta(A_1) \neq f^\beta(A_2)$. Then, as above, $f^\alpha \circ \pi_\alpha(A_1) = f^\beta(A_1) \neq f^\beta(A_2) = f^\alpha \circ \pi_\alpha(A_2)$ so that $\pi_\alpha(A_1) \neq \pi_\alpha(A_2)$.

The above shows that $\alpha X \setminus X$ is infinite.

We must finally prove that $\alpha X \setminus X$ is totally disconnected. Let $a \in \alpha X \setminus X$, let O be an open set in $\alpha X \setminus X$ containing a, and let A be the component of $\beta X \setminus X$ for which $\pi_\alpha(A) = a$. $\pi_\alpha^{-1}(O)$ is an open set in $\beta X \setminus X$ containing A. By 6.30 A is the intersection of all open and closed sets in $\beta X \setminus X$ which contain it. These sets will be compact. As we saw in the proof of 6.29 there will be a finite collection $O_1, O_2, \ldots, O_n$ of open and closed sets such that $A \subset O_1 \cap \ldots \cap O_n \subset \pi_\alpha^{-1}(O)$. Let $V = O_1 \cap \ldots \cap O_n$. Then V is open and closed so that $\pi_\alpha(V)$ is open (π_α is open and takes points of $\beta X \setminus X$ into $\alpha X \setminus X$) and closed (V is compact) in $\alpha X \setminus X$. Thus,

$$a = \pi_\alpha(A) \in \pi_\alpha(V) \subset O.$$

This shows that $\alpha X \setminus X$ has a base of open and closed sets and so, by 6.31, is totally disconnected.

(b) implies (c): By 6.31 $\alpha X \setminus X$ has a base of open and closed sets. Thus there is a family of non-empty, pairwise disjoint, open and closed sets $\{F_n\}_{n=1}^\infty$. Let $F_o = (\alpha X \setminus X) \setminus \bigcup_{n=1}^\infty F_n$ and let $F = \{f \in C_\alpha | f^\alpha$ is constant on each F_n, $n = 1, 2, \ldots\}$. Let $\gamma X = e_F X$. It is easily verified that $|\gamma X \setminus X| = \aleph_0$. ($\alpha X \geq \gamma X$ and the quotient map sends each F_i onto a separate point of $\gamma X \setminus X$.)

(c) implies (d): $\gamma X \setminus X$ must be totally disconnected since it is homeomorphic to a countable subset of $\mathbb{R}$ (5.40). By 6.31 we may obtain n non-empty, pairwise disjoint, open and closed sets in $\gamma X \setminus X$ whose union is $\gamma X \setminus X$, say $F_1, \ldots, F_n$. (As above, obtain $n - 1$ of them and let $F_n = (\gamma X \setminus X) \setminus (F_1 \cup \ldots \cup F_{n-1})$. Then F_n is open and closed).

Let $\mathcal{F} = \{f \in \mathbf{C}_\gamma | f^\gamma$ is constant on each F_i, $i = 1,\ldots,n\}$. Let $\alpha_n X = e_{\mathcal{F}} X$.

(d) implies (a): If X has an n-point compactification $\alpha_n X$ for each $n \geq 1$ then X is locally compact and $\beta X \setminus X$ must have n components for each $n \geq 1$ since there is a map from $\beta X \setminus X$ onto $\alpha_n X \setminus X$. Thus $\beta X \setminus X$ has infinitely many components.

6.33 Remarks. We conclude this chapter with a nice result of Levy and McDowell [1975] concerning a cardinal-topological property of remainders which therefore can act as a natural bridge to the next chapter (which is concerned with topological properties of the remainder).

6.34 Lemma. For $\alpha X \in K(X)$ we have D dense in βX if and only if $\pi_\alpha(D)$ is dense in αX.

Proof. If $\overline{D} = \beta X$ then $\alpha X = \pi_\alpha(\beta X) = \pi_\alpha(\overline{D}) \subset \overline{\pi_\alpha(D)} \subset \alpha X$.

Conversely, if $\pi_\alpha(D)$ is dense in αX then $\pi_\alpha(\overline{D})$ is a compact and thus closed subset of αX containing $\pi_\alpha(D)$. It follows that $\pi_\alpha(\overline{D}) = \alpha X$. Since π_α is the identity map on X and sends $\beta X \setminus X$ to $\alpha X \setminus X$, we must have $X \subset \overline{D} \subset \beta X$. Thus $\overline{D} = \beta X$; i.e., D is dense in βX.

6.35 Definitions. The density character of a topological space X is the least cardinal number of the dense subsets of X. A space is hereditarily separable if every subspace of it is separable.

6.36 Theorem. The following are equivalent:

(i) Every dense subset of βX has density character $\leq m$.

(ii) Every dense subset of αX has density character $\leq m$ for every $\alpha X \in K(X)$.

(iii) Every dense subset of αX has density character $\leq m$ for some $\alpha X \in K(X)$.

Proof. (i) implies (ii): If D is dense in αX then by 6.34 $\pi_\alpha^{-1}(D)$ is dense in βX and so has a dense subset D_o of cardinality $\leq m$. $\pi_\alpha(D_o)$ is a dense subset of αX with cardinality $\leq m$.

(ii) implies (iii): Obvious.

(iii) implies (i): If D is dense in βX then $\pi_\alpha(D)$ is dense in αX and so contains a dense subset D_o with $|D_o| \leq m$. For each $d \in D_o$ choose $x_d \in \pi_\alpha^{-1}(d) \cap D$. $\{x_d\}_{d \in D_o}$ will be a dense subset of D with cardinality $\leq m$.

<u>6.37 Corollary</u>. If X has no compact neighborhoods and has a compactification which is hereditarily separable then $\beta X \smallsetminus X$ is separable.

<u>Proof</u>. If $x \in \overline{\beta X \smallsetminus X}$ then $U = \beta X \smallsetminus (\overline{\beta X \smallsetminus X})$ contains a compact neighborhood of x and since $U \subset X$ this cannot be. Thus $\beta X \smallsetminus X$ is dense in βX. By 6.36 (with $m = \aleph_0$) it follows that $\beta X \smallsetminus X$ is separable.

<u>6.38 Corollary</u>. All remainders of Q are separable. Likewise, all remainders of $J = \mathbb{R} \smallsetminus Q$ are separable.

<u>Proof</u>. 6.37 and 6.34.

<u>6.39 Major Problem #2</u>. The major portion of this chapter has been concerned with questions concerning the cardinality of remainders. The finite and countable cases have been well-studied and nice results are available. Remainders with cardinality between $\aleph_0$ and 2^c do not seem to have been generally considered at all. Thus, Major Problem #2 is to obtain necessary and/or sufficient conditions on X to guarantee that X has remainders with specific cardinality between $\aleph_0$ and 2^c.

§7: REMAINDERS - CONNECTED AND DISCONNECTED

7.1 Remarks. In this chapter we will be concerned with some specific aspects of the following general question: For a given space X, which spaces can be remainders of X? (Note: At present, there seems to be no concensus as to what to call $\alpha X \setminus X$. The terms in common usage today are "remainders", "outgrowth", and, at least in the case of $\beta X \setminus X$, "infinity". This last seems to be a poor choice compared to the first two. In talking about X, the term "outgrowth" seems preferable, whereas when referring to αX, "remainder" seems more logical.) We first discuss several miscellaneous results, all concerned with conditions which guarantee that the remainders will be connected. The first result, however, is more general.

7.2 Theorem (Magill [1966]). For a locally compact space X the following are equivalent.

(a) Y is a remainder of X.

(b) Y is the continuous image of a remainder of X.

(c) Y is the continuous image of $\beta X \setminus X$.

Proof. (a) implies (b) is trivial and since any remainder is a continuous image of $\beta X \setminus X$ (2.8, 1.30), (c) follows from (b). Hence we need to prove only that (c) implies (a). Let $g: \beta X \setminus X \to Y$ (onto) and let

$$\mathcal{F} = \{f \in C^*(X) \mid f^\beta|_{g^{-1}(p)} \text{ is constant for each } p \in Y\}.$$

$\mathcal{F}$ separates points and closed sets in X since X is locally compact. (The subset $\mathcal{F}_o \subset \mathcal{F}$ consisting of those f for which f^β is constant on all of $\beta X \setminus X$ separates points and closed sets in X.) Thus, $\mathcal{F}$ determines a compactification $\alpha X = e_{\mathcal{F}} X$.

Suppose $r, s \in g^{-1}(p)$. Then for each $f \in \mathcal{F}$ we have $f^\beta(r) = f^\beta(s)$ so that

$$f^\alpha \circ \pi_\alpha(r) = f^\beta(r) = f^\beta(s) = f^\alpha \circ \pi_\alpha(s)$$

and it follows that $\pi_\alpha(r) = \pi_\alpha(s)$. Thus, $\pi_\alpha(g^{-1}(p))$ is a point of $\alpha X \setminus X$ for each $p \in Y$. Define $h: \alpha X \setminus X \to Y$ by

$$h(\pi_\alpha(g^{-1}(p))) = p.$$

h is well-defined since each point of $\alpha X \setminus X$ is the image under π_α of a point of $\beta X \setminus X$ and if two points of $\beta X \setminus X$ go to the same point of $\alpha X \setminus X$ then they must be in the same $g^{-1}(p)$. (Otherwise, some $f \in F$ would distinguish between them and hence they would go to distinct points under π_α.)

For $u, v \in \alpha X \setminus X$ we have $u = \pi_\alpha(g^{-1}(p))$ and $v = \pi_\alpha(g^{-1}(q))$ for some $p, q \in Y$. $h(u) = h(v)$ implies $p = q$ which implies that $g^{-1}(p) = g^{-1}(q)$ so that $u = v$. This says that h is $1 - 1$. To see that h is continuous, observe that for a closed $U \subset Y$ we have $h^{-1}(U) = \pi_\alpha(g^{-1}(U))$ which is closed since g is continuous and π_α is closed. Therefore, h is a homeomorphism ($1 - 1$ and continuous from $\alpha X \setminus X$ (compact) onto Y (Hausdorff).)

<u>7.3 Example</u>. The purpose of this example is to show that local compactness is essential in the above theorem. In $\mathbb{R}^2$ let $L_n = \{(x,y) | x = 1/n,\ 0 \leq y \leq 1\}$, let $X = [0,2] \times [0,1] \setminus \bigcup_{n=1}^{\infty} L_n$ and let $\alpha X = [0,2] \times [0,1]$.

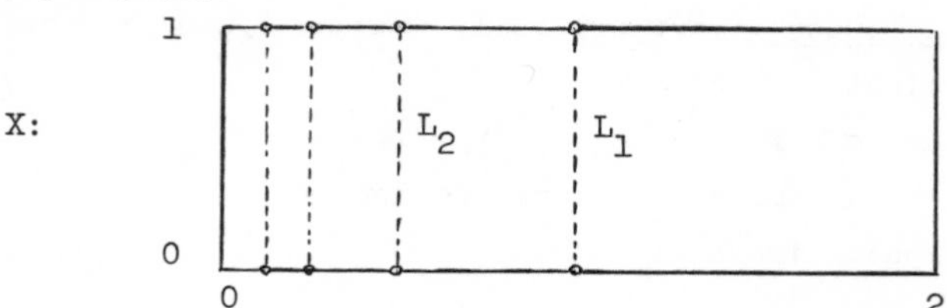

For $Y = \{(1/n, 0)\}_{n=1}^{\infty} \subset \mathbb{R}^2$ we have a map $g\colon \alpha X \setminus X \to Y$ defined by $g(L_n) = (1/n, 0)$ (note that $\alpha X \setminus X = \bigcup_{n=1}^{\infty} L_n$). g is a continuous, open, perfect map (perfect = $g^{-1}(p)$ is compact for each $p \in Y$) but no (Hausdorff) compactification of X exists whose remainder is homeomorphic to Y. We had originally conjectured an analogue to 7.2 for X not necessarily locally compact provided we hypothesied that (i) $\alpha X \setminus X = \bigcup_{n=1}^{\infty} g^{-1}(p_n)$, (ii) g was continuous, open, and perfect. This example annihilated such hopes.

<u>7.4 Definitions</u>. A <u>continuum</u> is a compact connected Hausdorff space. A <u>Peano space</u> is any Hausdorff space which is the continuous image of $[0,1]$. The famous Hahn-Mazurkiewicz theorem (Hahn [1914], Mazurkiewicz [1920]) characterizes the Peano spaces as locally connected, metric continua. A <u>weak Peano space</u> is any compact Hausdorff space which

contains a dense, continuous image of $\mathbb{R}$. An example of a weak Peano space which is not a Peano space is the closure in $\mathbb{R}^2$ of the graph of $y = \sin 1/x$, $0 < x \leq 1$.

7.5 Corollary. If X has a remainder which is a non-degenerate (at least two points) metric continuum Y, then any Peano space is a remainder of X.

Proof. All we need show is that $[0,1]$ is a continuous image of Y. The corollary will then follow from 7.2. Define $f: Y \to [0,1]$ by $f(x) = d(x,a)/[d(x,a) + d(x,b)]$, where d is the metric on Y, and a and b are two distinct points. We have that $f(a) = 0$, $f(b) = 1$, $0 \leq f(x) \leq 1$, and since $f(Y)$ is connected, $f(Y) = [0,1]$.

7.6 Lemma (Steiner and Steiner [1968]). Let X be locally compact and non-compact and let K be compact. If there is a map $f: X \to K$ such that $f(X \setminus F)$ is dense in K for each compact $F \subset X$ then K is a remainder of X.

Proof. Use the construction of 2.12 to obtain a compactification $\alpha X = \overline{c_{\mathcal{F}}(X)} \subset \omega X \times Y$ where $\mathcal{F} = \{\omega, f\}$ and $c_{\mathcal{F}}: X \to \omega X \times Y$ is the evaluation map. Now no point of the form (x,k) where $k \neq f(x)$ is in $\overline{c_{\mathcal{F}}(X)}$ since f is continuous and K is Hausdorff. Also, every point of the form (∞,k), $k \in K$ is in $\overline{c_{\mathcal{F}}(X)}$ since the neighborhoods of ∞ in ωX are precisely the complements of the compact sets in X, every one of which is mapped densely into K. Thus, $\alpha X \setminus X = \{\infty\} \times K$ which is homeomorphic to K.

7.7 Remark. We used a different construction in 4.20 to obtain the "large" compactification of N. An easy consequence of the technique of construction in 4.20 is that any compact separable space is a remainder of N. This can be generalized to the following. Let D be an infinite discrete space and let Y be a compact space with a dense subset whose cardinality is less than or equal to $|D|$. Then Y is a remainder of D.

7.8 Theorem. If X is locally compact and non-pseudocompact and Y is any weak Peano space, then Y is a remainder of X.

Proof. By 6.22 X contains a subspace S which is C-embedded and homeomorphic to N. Let $g: S \to Q$ (= rationals) be any onto mapping. This extends to $h: X \to \mathbb{R}$. If $\overline{D} = Y$ where $t: \mathbb{R} \to D$ (onto) is continuous, let $f = t \circ h$. For F compact in X, we have that F

contains only finitely many points of S. Thus, $h(X\setminus F)$ is dense in $\mathbb{R}$ so that $f(X\setminus F)$ is dense in Y. By 7.6, the theorem follows.

7.9 Corollary (Magill [1966]). If X is a locally compact, normal space which contains an infinite, discrete, closed subspace and Y is any continuous image of $[0,1]$ then Y is a remainder of X.

7.10 Corollary (Rogers [1969]). If X is locally compact and non-pseudocompact then any Peano space is a remainder of X.

7.11 Corollary (Magill [1970]). Suppose X is locally compact and contains a connected, C-embedded subset which is not pseudocompact and Y contains a dense subspace which is arcwise connected and separable. Then each compactification of Y is a remainder of X.

Proof. The hypothesis on X assures us that X is not pseudocompact. The hypothesis on Y assures us that Y contains a dense continuous image of $\mathbb{R}$ so that each compactification of Y is a weak Peano space. The result thus follows from 7.8. (Since Y is separable, let $\{y_n\}_{n=-\infty}^{\infty}$ be a countable dense subset of Y. Define $f\colon \mathbb{R} \to Y$ as follows

$$f(r) = h_n(r), \quad \text{where} \quad n \leq r \leq n + 1$$

and $h_n\colon [n, n + 1] \to Y$ is an arc from y_n to y_{n+1}, $-\infty < n < +\infty$.)

7.12 Remarks. If X is not locally compact then none of the preceeding results apply. In particular, 7.2 is false: let Y be a single point. Then (b) and (c) hold but (a) does not. Also, no remainder of X can be compact and so no continua will be a remainder. Remainders of non-locally compact spaces can be connected: Let $X = [0,1] \times [0,1]\setminus\{1\} \times (0,1)$. For $\alpha X = [0,1] \times [0,1]$ we have $\alpha X\setminus X = \{1\} \times (0,1)$ which is connected:

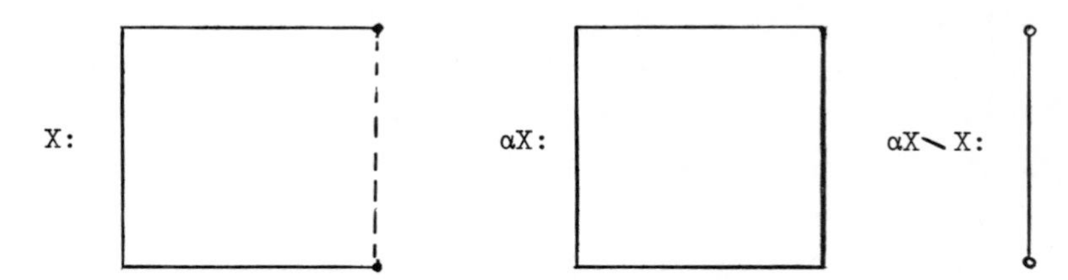

Recall that $K(f)$ is the set of elements of $K(X)$ to which $f \in C^*(X)$ has an extension (5.33). We can say something about the minimal elements of $K(f)$ if $\beta X\setminus X$ is a continuum:

7.13 Theorem. If $\beta X\setminus X$ is compact and connected then if αX is a minimal element of $K(f)$ either $\alpha X\setminus X$ is a single point or is homeomorphic to $[0,1]$.

<u>Proof</u>. By 5.29 f^α is $1-1$ on $\alpha X \setminus X$. Since $\alpha X \setminus X$ is compact and connected f^α is a homeomorphism. Since the only compact connected subsets of $\mathbb{R}$ are the singletons and the closed intervals, the theorem follows.

We will next consider an almost opposite type of question: When is a remainder totally disconnected? Theorem 6.32 gave equivalent conditions for a locally compact space to have an infinite, totally disconnected remainder (note that any finite remainder or any countable, compact remainder is totally disconnected.)

Our first result, due to Banaschewski [1974], is extremely general. We will need a few definitions and one lemma first.

<u>7.14 Definitions</u>. Let $\mathcal{H}$ be the class of Hausdorff spaces and let $\mathcal{A} \subset \mathcal{H}$ be a productive and closed-hereditary subclass (i.e., products and closed subsets of elements of $\mathcal{A}$ are members of $\mathcal{A}$.) Let $\mathcal{C} \subset \mathcal{H}$ be any productive and hereditary subclass. For a space X we say that E is an <u>extension</u> of X if X can be considered as a dense subset of E. <u>An $\mathcal{A}$-extension of X with remainder in $\mathcal{C}$</u> is any element E of $\mathcal{A}$ which is an extension of X for which $E \setminus X \in \mathcal{C}$. If E_1 and E_2 are extensions of X, we say $E_1 \geq E_2$ if there is a continuous map $f\colon E_1 \to E_2$ which leaves X pointwise fixed. E_1 and E_2 are <u>equivalent</u> if f is a homeomorphism.

<u>7.15 Lemma</u>. Let $\{E_\alpha\}_{\alpha\in\Gamma}$ be a family of extensions of X and suppose $\mathcal{F} = \{i_\alpha\colon X \to E_\alpha\}_{\alpha\in\Gamma}$ is the family of inclusion mappings. A point p in the closure of $e_{\mathcal{F}}(X)$ in $\prod_{\alpha\in\Gamma} E_\alpha$ which has at least one projection, say π_{α_0}, in $i_{\alpha_0}(X)$ is in $e_{\mathcal{F}}(X)$.

<u>Proof</u>. There is a net $\{x_\lambda\}_{\lambda\in\Lambda}$ in X for which $e_{\mathcal{F}}(x_\lambda) \to p$. Then

$$i_{\alpha_0}(x_\lambda) = \pi_{\alpha_0}(e_{\mathcal{F}}(x_\lambda)) \to \pi_{\alpha_0}(p) = i_{\alpha_0}(x)$$

for some $x \in X$. Thus in X, $x_\lambda \to x$. For any $\alpha \in \Gamma$ we therefore have

$$\pi_\alpha(e_{\mathcal{F}}(x_\lambda)) = i_\alpha(x_\lambda) \to i_\alpha(x)$$

so that $\pi_\alpha(p) = i_\alpha(x)$ for all α. It follows that $p \in e_{\mathcal{F}}(X)$.

<u>7.16 Theorem</u>. If X has any $\mathcal{A}$ extensions with remainder in $\mathcal{C}$, then it has a maximal one (with respect to $\geq$).

Proof. Let $\{E_\alpha\}_{\alpha\varepsilon\Gamma}$ be a family of inequivalent A extensions with remainders in C having the property that any A extension with remainder in C is equivalent to some E_α. An argument entirely analogous to that in 3.22 will establish that Γ is a set. By 7.15, the remainder $\overline{e_F(X)} \smallsetminus e_F(X)$ can be embedded in $\prod_{\alpha\varepsilon\Gamma}(E_\alpha \smallsetminus X)$. Thus, $E = \overline{e_F(X)}$ is an A extension with remainder in C. The projection maps, restricted to $E \subset \prod_{\alpha\varepsilon\Gamma} E_\alpha$, establish that $E \geq E_\alpha$ for all $\alpha \varepsilon \Gamma$.

7.17 Remarks. For A the subclass of compact (Hausdorff) spaces, an A extension will be simply a compactification. Thus for any property P which is productive and hereditary and for which X has a compactification whose remainder has property P, X has a maximal compactification whose remainder has property P. Unfortunately, few topological properties are both productive and hereditary. However, total disconnectedness is one such.

7.18 Corollary. If X has a compactification with totally disconnected remainder, it has a maximal one.

7.19 Definition. The maximal compactification with totally disconnected remainder is called the Stoilow-Kerékjárto compactification of X and is denoted by κX.

7.20 Corollary. If X is locally compact, κX exists.

Proof. In this case $\omega X \smallsetminus X = \{\infty\}$ is totally disconnected.

7.21 Lemma (McCartney [1971]). Let X be locally compact. If $\alpha X \smallsetminus X$ is totally disconnected and has more than one point, then αX is the least upper bound of all two point compactifications less than or equal to αX.

Proof. Let $\{\tau_i X\}_{i\varepsilon I}$ be the set of two point compactifications less than or equal to αX. Clearly $\alpha X \geq \text{lub}\{\tau_i X\}_{i\varepsilon I}$. If p_1 and p_2 are distinct points of $\alpha X \smallsetminus X$ then we may partition $\alpha X \smallsetminus X$ into open and closed sets A_1, A_2 such that $p_i \varepsilon A_i$, $i = 1,2$. The quotient space obtained by collapsing A_1 and A_2 to points is equal to $\tau_i X$ for some $i \varepsilon I$. For that i we have $\alpha X \geq \text{lub}\{\tau_i X\}_{i\varepsilon I} \geq \tau_i X$ and so the quotient map from αX to $\tau_i X$ factors through $\text{lub}\{\tau_i X\}_{i\varepsilon I}$. As the first quotient map distinguishes between p_1 and p_2, so must the second. Since p_1 and p_2 were arbitrary points of $\alpha X \smallsetminus X$, the quotient map of αX onto $\text{lub}\{\tau_i X\}_{i\varepsilon I}$ is 1 - 1 and hence is a homeomorphism.

7.22 Theorem (McCartney [1971]). For X locally compact if κX exists and is not equal to ωX, then κX is the least upper bound of all two point compactifications of X.

7.23 Corollary. $\kappa\mathbb{R}^n = \omega\mathbb{R}^n$ if $n > 1$. $\kappa\mathbb{R} = \tau\mathbb{R}$ (the two point compactification of $\mathbb{R}$).

7.24 Definition. A space X is zero dimensional if each point has a base of open and closed sets. This is easily seen to be a productive, hereditary property.

7.25 Theorem. If X has a compactification with zero dimensional remainder then it has a maximal one.

Proof. Immediate from 7.16.

7.26 Definition. The maximal compactification with zero dimensional remainder is called the Freudenthal compactification (Freudenthal [1942]) and is denoted by ϕX.

7.27 Corollary. If X is locally compact, ϕX exists.

7.28 Remarks. We end this chapter with a result which, in some sense, is a unifying result giving a sufficient condition for a space to have both connected and disconnected remainders; actually it will have all compact metric spaces for remainders. In order to obtain this result we need several incidental (but difficult) results.

7.29 Theorem (Urysohn [1925]). A regular space X satisfying the second axiom of countability is metrizable.

Proof. Let $\mathcal{B}$ be a countable base for X and let

$$\mathcal{A} = \{(U,V) \mid U,V \in \mathcal{B} \text{ and } \overline{U} \subset V\}.$$

$\mathcal{A}$ is countable. Since X is regular and Lindelöf it is normal (4.38) so that for each $(U,V) \in \mathcal{A}$ there is a function $f_{U,V}\colon X \to [0,1]$ such that $f(\overline{U}) = 0$ and $f(X \setminus V) = 1$ (1.8). Let $\mathcal{F} = \{f_{U,V} \mid (U,V) \in \mathcal{A}\}$. We claim $\mathcal{F}$ separates points from closed sets in X. If F is a closed set and $x \in X \setminus F$ then there is an element $V \in \mathcal{B}$ with $x \in V \subset X \setminus F$. Since X is regular there is an open set O such that $x \in O \subset \overline{O} \subset V$. There is a $U \in \mathcal{B}$ with $x \in U \subset O$. Clearly $(U,V) \in \mathcal{A}$ and $f_{U,V}$ distinguishes x and F. Let $e_{\mathcal{F}}\colon X \to \prod_{(U,V)\in\mathcal{A}} I_{(U,V)}$ (where $I_{(U,V)} = [0,1]$) be the diagonal map (1.23). By 1.24 this is an embedding and since the product $\prod_{(U,V)\in\mathcal{A}} I_{(U,V)}$ is a countable product of metric spaces it is metrizable. (If $x = (x_1, x_2, \ldots)$ and $y = (y_1, y_2, \ldots)$ define $d(x,y) = \sum_{n=1}^{\infty} 2^{-n}|x_n - y_n|$.)

7.30 Corollary. If X is a compact metric space and Y is Hausdorff and if $f\colon X \to Y$ is continuous then $f(X)$ is compact and metrizable.

Proof. In view of 7.29, all we need to show is that $f(X)$ satisfies the second axiom of countability. Let $\mathcal{B}$ be a countable base for X. (Note that compact metric spaces satisfy the second axiom of countability: for each n there is a finite open cover of X with sets of diameter $1/n$. The union of all such covers is a countable base.) Let $\mathcal{A}$ be the collection of all finite unions of elements from $\mathcal{B}$ and let $\mathcal{C} = \{f(X) \setminus f(X \setminus A) | A \in \mathcal{A}\}$. We claim $\mathcal{C}$ is a countable base for $f(X)$. (A is open so $X \setminus A$ is compact. Thus, $f(X \setminus A)$ is closed in Y and therefore $f(X) \setminus f(X \setminus A)$ is open in $f(X)$.) Suppose 0 is open in $f(X)$ and $p \in 0$. Then $f^{-1}(p)$ is a closed and hence compact subset of $f^{-1}(0)$. Thus there are sets $0_1, \ldots, 0_n \in \mathcal{B}$ with $f^{-1}(p) \subset 0_1 \cup \ldots \cup 0_n \subset f^{-1}(0)$. If $U = 0_1 \cup \ldots \cup 0_n$ then $U \in \mathcal{A}$ and $p \in f(X) \setminus f(X \setminus U)$.

7.31 Definition. A Cantor set is any countably infinite product of finite, non-singleton, discrete spaces. It is well-known that all Cantor sets are homeomorphic. (For example, see Willard [1968], Chapter 30.) We do not prove this here. We will prove, however, the celebrated result of Alexandroff-Urysohn [1929], following the proof by Schoenfield [1974].

7.32 Theorem. Any compact metric space is the continuous image of a Cantor set.

Proof. We construct a sequence $\{\mathcal{U}_n\}_{n=1}^{\infty}$ of covers of X inductively as follows:

For $n = 1$, $\mathcal{U}_1$ is a collection $U_1, U_2, \ldots, U_{N1}$ of sets which are closures of non-empty open sets, the diameter of U_i is less than 2^{-1}, and $N1 > 1$. Suppose $\mathcal{U}_1, \mathcal{U}_2, \ldots, \mathcal{U}_n$ have been chosen satisfying

(i) Each $U \in \mathcal{U}_i$ is the closure of a non-empty open set in X of diameter less than 2^{-i}.

(ii) If $\mathcal{U}_i = \{U_{j1,j2,\ldots,ji} \mid 1 \leq jk \leq N_k$ for $1 \leq k \leq i\}$ then $N_k > 1$ for each k and for fixed $j1, j2, \ldots, ji$ we have

$$U_{j1,j2,\ldots,ji} = \bigcup_{j(i+1)=1}^{N_{i+1}} U_{j1,j2,\ldots,ji,j(i+1)}.$$

We obtain $\mathcal{U}_{k+1}$ by fixing an index $j1, j2, \ldots, jk$, covering the closed and hence compact set $U_{j1,j2,\ldots,jk}$ by open sets $\{O_{j1,j2,\ldots,jk,i}\}_{i=1}^{N_{j1,j2,\ldots,jk}}$ of diameter less than $2^{-(k+1)}$ such that for each i, $(O_{j1,j2,\ldots,jk,i}) \cap (U_{j1,j2,\ldots,jk}) \neq \phi$. Let $U_{j1,j2,\ldots,jk,i}$ be the closure of this intersection. Let $N_{k+1} = \max\{N_{j1,j2,\ldots,jk} \mid j1,j2,\ldots,jk$ is an index in $\mathcal{U}_k\}$. Then if, for example, N_{k+1} is greater than $N_{j1,j2,\ldots,jk}$ for a fixed index $j1,\ldots,jk$, let $U_{j1,j2,\ldots,jk,i} = U_{j1,j2,\ldots,jk,jN_{j1,\ldots,jk}}$ for $N_{j1,\ldots,jk} \leq i \leq N_{k+1}$. Then conditions (i) and (ii) above are satisfied for $\mathcal{U}_{k+1}$.

Let $D_n = \{1, 2, \ldots, N_n\}$ with the discrete topology and let $C = \prod_{n=1}^{\infty} D_n$. Then C is a Cantor set (by definition) and we may define $f: C \to X$ as follows:

$$f((i1, i2, \ldots)) = U_{i1} \cap U_{i1,i2} \cap U_{i1,i2,i3} \cap \cdots$$

By the Cantor intersection theorem $f((i1, i2, \ldots))$ is a uniquely determined point. (The intersection is of a nested family of closed sets whose diameters are decreasing to zero.) Since every point of X is determined by some sequence $(i1, i2, \ldots)$ (each $\mathcal{U}_n$ is a cover of X) it follows that f is onto. We show that f is continuous. Let $x = f((i1, i2, \ldots))$ and suppose O is a neighborhood of x. Since the diameter of the sets in $\mathcal{U}_n$ is less than 2^{-n} there is an N such that $U_{i1,i2,\ldots,iN} \subset O$. Let $U = \{i1\} \times \{i2\} \times \cdots \times \{iN\} \times \prod_{n=N+1}^{\infty} D_n$. Then U is open in C and $f(U) \subset O$. Thus f is continuous at an arbitrary point of C and hence is continuous.

<u>7.33 Definition</u>. For each $n \in N$, let $N_n^* = W(\omega + 1)$. (See 5.1.) Thus, N_n^* is simply the one-point compactification of N, using ω as the "point of infinity". Let $K = \prod_{n=1}^{\infty} N_n^*$.

<u>7.34 Lemma</u>. Any Cantor set is the continuous image of K.

<u>Proof</u>. If $C = \prod_{n=1}^{\infty} D_n$ where $D_n = \{1, 2, \ldots, k_n\}$ has the discrete topology, define $f_n: N_n^* \to D_n$ by

$$f_n(i) = \begin{cases} i, & \text{if } i \leq k_n \\ k_n, & \text{if } i > k_n \text{ or } i = \omega. \end{cases}$$

Clearly f_n is continuous for each n and if we define $f\colon K \to C$ by

$$f((i_1, i_2, \ldots)) = (f_1(i_1), f_2(i_2), \ldots)$$

then f is continuous and onto.

<u>7.35 Theorem</u>. If $\beta X \setminus X = K$ then every remainder of X is a compact metric space and every compact metric space is a remainder of X.

<u>Proof</u>. First, by 7.2 and 7.30, any remainder of X is a compact metric space. In view of 7.2 and 7.34 all Cantor spaces are remainders of X. Finally, by 7.2 and 7.32, any compact metric space is a remainder of X.

<u>7.36 Example</u>. Let X be a space for which $\beta X \setminus X = K \times I$ (4.17). K is a continuous image of $\beta X \setminus X$ (via the projection map) so by 7.2 $K = \alpha X \setminus X$ for some $\alpha X \in K(X)$. By the Alexandroff-Urysohn theorem (7.32) and Lemma 7.34 $K \times I$ is a continuous image of K. So by 7.2 again, $K \times I = \gamma X \setminus X$ for some $\gamma X \in K(X)$. By examining the proof of 7.2 we see that $\beta X > \alpha X > \gamma X$. Thus we have the seemingly rather bizarre situation where $\beta X > \gamma X$ but $\beta X \setminus X = K \times I = \gamma X \setminus X$. Actually, it is quite easy to obtain examples where $\alpha X > \gamma X$ but $\alpha X \setminus X = \gamma X \setminus X$ (e.g., if $\alpha X \setminus X = [0,1]$, obtain γX as a quotient space of αX by making all points of $[1/3, 2/3]$ equivalent).

<u>7.37 Major Problem #3</u>. The result 7.8 seems to be about the best available giving conditions on X which guarantee some compact, connected space is a remainder of X. It leaves a lot to be desired: The subset of the plane consisting of the graph of $y = \sin 1/x$, $0 < x \leq 1$, together with the interval $-2 \leq y \leq 2$, $x = 0$, is clearly a remainder of $\mathbb{R}$. (Draw $\mathbb{R}$ in the plane with its two ends "spiralling" down on this set.) However, it is <u>not</u> a weak Peano space so that 7.8 does not apply. Thus, Major Problem #3 is to obtain a more general result than 7.8; one that includes a larger class of continua than weak Peano spaces.

<u>7.38 Major Problem #4</u>. Almost all results on remainders are for locally compact spaces. One major source for the non-locally compact case is the paper by Rayburn [1973]. We will cite two of his theorems, both of which are generalizations of results of Magill which we have discussed previously (7.2 and 5.27). $R(X)$ denotes the points of X which do not have any compact neighborhoods (in X). For any $\alpha X \in K(X)$ we have $R(X) = X \cap \overline{\alpha X \setminus X}$.

Theorem 1.1. For spaces X and Y there is a compactification αX with Y homeomorphic to $\overline{\alpha X \setminus X}$ if and only if there is continuous map of $\overline{\beta X \setminus X}$ onto Y which is $1-1$ on $R(X)$.

Theorem 2.2. For spaces X and Y if there is a homomorphism from $\overline{\beta X \setminus X}$ onto $\overline{\beta Y \setminus Y}$ taking $R(X)$ onto $R(Y)$ then $K(X)$ and $K(Y)$ are lattice isomorphic.

Note that these results are not really about remainders but about closures of remainders. Nothing seems to be known about remainders. Specifically, we can easily give examples of spaces which are remainders of Q, say, (i.e., $[0,1] \setminus Q$, since $Q \cap [0,1]$ is homeomorphic to Q) and we can easily give examples of spaces which are not remainders of Q (i.e., Y for any compact Y). However, we know of no general criteria which, if met by a space Y, will guarantee that Y is a remainder of Q, nor do we know a general criteria (other than non-compactness) which all remainders of Q satisfy except for the Levy-McDowell result 6.38 that they are separable.

Generally then, Major Problem #4 could be to study the topological properties of the remainders for non-locally compact spaces and to give a "large" class of spaces which are remainders for some other "large" class of spaces.

§8: WALLMAN-FRINK COMPACTIFICATIONS

Wallman [1938] gave a general method for constructing a T_1-compactification for any T_1-space X. If X were normal then this compactification turned out to be βX. Frink [1964] showed how to modify Wallman's method so as to obtain a whole family of Hausdorff compactifications of a completely regular space X. Using the development of Gillman and Jerison [1960] which produces the compactification σX (3.11) and which had been derived from the constructions of Samuel [1948] and ultimately from Wallman's method, Frink abstracted those properties which a family $\mathcal{C}$ of closed subsets of X needed so as to produce a Hausdorff compactification by the method of topologizing the set of $\mathcal{C}$-ultrafilters.

<u>8.1 Definition</u>. A family $\mathcal{C}$ of closed subsets of X is a <u>normal base</u> for X if

(i) $C_1, C_2 \in \mathcal{C}$ imply $C_1 \cup C_2 \in \mathcal{C}$ and $C_1 \cap C_2 \in \mathcal{C}$. ($\mathcal{C}$ is a <u>ring</u>.)

(ii) If F is closed in X and $x \in X \setminus F$ then there is a $C \in \mathcal{C}$ with $x \in C$ and $C \cap F = \emptyset$.

(iii) If $A_1, A_2 \in \mathcal{C}$ with $A_1 \cap A_2 = \emptyset$ then there exist $C_1, C_2 \in \mathcal{C}$ with $A_1 \cap C_1 = \emptyset$, $A_2 \cap C_2 = \emptyset$, and $C_1 \cup C_2 = X$.

(iv) $\mathcal{C}$ is a base for the closed sets in X.

<u>8.2 Construction</u>. We proceed exactly analogously to 3.11: Let $\omega_{\mathcal{C}}X$ be the set of $\mathcal{C}$-ultrafilters, i.e., the set of families of subsets of $\mathcal{C}$ (i) which are closed under finite intersections, (ii) contain all $\mathcal{C}$-"supersets" of their members, and (iii) are maximal with respect to properties (i) and (ii). (The construction of 3.11 used $\mathcal{C} = Z[X]$, the zero sets of X, which is a normal base, as Frink observed.)

We give $\omega_{\mathcal{C}}X$ a topology by taking as a base for the closed sets all sets of the form $\overline{C}$, $C \in \mathcal{C}$, where $\overline{C}$ is the set of all $\mathcal{C}$-ultrafilters which contain C. Then $\omega_{\mathcal{C}}X$ is compact (since the $\mathcal{C}$-ultrafilters have the finite intersection property), X is embedded densely in $\omega_{\mathcal{C}}X$ (a point $x \in X$ is identified with the $\mathcal{C}$-ultrafilter fixed at x, namely the collection of all elements of $\mathcal{C}$ which contain x), and

$\omega_C X$ is Hausdorff (by using property (iii) of the definition of normal base).

8.3 Theorem. For $\alpha X \in K(X)$ the following are equivalent:

(1) There is a normal base $\mathcal{C}$ for X such that $\alpha X \approx \omega_C X$.

(2) There is a base ring $\mathcal{A}$ of closed sets in αX such that $\overline{A \cap X} = A$ for all $A \in \mathcal{A}$.

(3) There is a base ring $\mathcal{B}$ of closed sets in αX such that $B \in \mathcal{B}$, $B \neq \emptyset$ implies that $B \cap X \neq \emptyset$.

Proof. (1) implies (2): Let $\mathcal{A} = \{\overline{C} | C \in \mathcal{C}\}$. Then $\overline{C} \cap X = C$ so that $\overline{\overline{C} \cap X} = \overline{C}$.

(2) implies (3): Let $\mathcal{B} = \mathcal{A}$. Then $B \in \mathcal{B}$, $B \neq \emptyset$, implies that $\overline{B \cap X} = B \neq \emptyset$ so $B \cap X \neq \emptyset$.

(3) implies (1): Let $\mathcal{C} = \{B \cap X | B \in \mathcal{B}\}$. To show $\mathcal{C}$ is a normal base for X we must verify that the four properties of the definition 8.1 are satisfied. The only one which is not obvious is property (iii). This follows from the fact that αX, a compact Hausdorff space, is normal: Suppose $A_1 = B_1 \cap X$, $A_2 = B_2 \cap X$ and $A_1 \cap A_2 = \emptyset$. Then $B_1 \cap B_2 = \emptyset$, for if $B_1 \cap B_2 \neq \emptyset$ we have $B_1 \cap B_2 \in \mathcal{B}$ so that $(B_1 \cap B_2) \cap X \neq \emptyset$, by hypothesis. Because αX is normal, there are disjoint open sets 0_1, 0_2 with $B_1 \subset 0_1$, $B_2 \subset 0_2$. B_1 is compact, $\alpha X \setminus 0_1$ is closed and disjoint from B_1, and $\mathcal{B}$ is a base for the closed sets in αX. Thus we may obtain a closed set $C_1' \in \mathcal{B}$ such that $C_1' \cap B_1 = \emptyset$ and $\alpha X \setminus 0_1 \subset C_1'$. Similarly, we may obtain a closed set $C_2' \in \mathcal{B}$ with $C_2' \cap B_2 = \emptyset$ and $\alpha X \setminus 0_2 \subset C_2'$. Let $C_1 = C_1' \cap X$, $C_2 = C_2' \cap X$. Then $A_1 \cap C_1 = \emptyset$, $A_2 \cap C_2 = \emptyset$, and $C_1 \cup C_2 = X$.

We must now show that $\alpha X \approx \omega_C X$. Define $\phi\colon \alpha X \to \omega_C X$ as follows:

$$\phi(p) = \{B \cap X \in \mathcal{C} | p \in B\}.$$

It is completely routine to verify that $\phi(p)$ is a $\mathcal{C}$-filter. We show it is maximal: Suppose $\phi(p) \subset \mathcal{S}$, a $\mathcal{C}$-filter. If $F \cap X \in \mathcal{S}$ and $F \cap X \notin \phi(p)$ for some $F \in \mathcal{B}$ then $p \notin F$ so that there exists $G \in \mathcal{B}$ with $p \in G$ and $G \cap F = \emptyset$ since $\mathcal{B}$ is a base. But then $G \cap X \in \phi(p)$ and $(G \cap X) \cap (F \cap X) = \emptyset$ so we see that $G \cap X \notin \mathcal{S}$, a contradiction. Thus, no $\mathcal{C}$-filter can properly contain $\phi(p)$ so that $\phi(p)$ is a $\mathcal{C}$-ultrafilter. In other words, $\phi(p) \in \omega_C X$. It is clear that ϕ maps $X \subset \alpha X$ identically onto $X \subset \omega_C X$.

ϕ is $1-1$ since if $p \neq q$ in αX then there is $B \in \mathcal{B}$ with $p \in B$ and $q \notin B$. It follows that $B \cap X \in \phi(p)$ and $B \cap X \notin \phi(q)$.

Thus, $\phi(p) \neq \phi(q)$.

ϕ is onto: If $\mathcal{F}$ is a $\mathcal{C}$-ultrafilter (i.e., $\mathcal{F} \in \omega_{\mathcal{C}}X$) then $\{F \in \mathcal{B} | F \cap X \in \mathcal{F}\}$ is a family of closed sets in αX with the finite intersection property. Because αX is compact, there is a point p in its intersection. Thus, $\mathcal{F} \subset \phi(p)$ and since $\mathcal{F}$ and $\phi(p)$ are $\mathcal{C}$-ultrafilters, we have $\mathcal{F} = \phi(p)$.

Finally, ϕ is continuous (and therefore is a homeomorphism since αX is compact and $\omega_{\mathcal{C}}X$ is Hausdorff) for if $\overline{C} = \overline{B \cap X}$ is a basic closed set in $\omega_{\mathcal{C}}X$, where $B \in \mathcal{B}$, we have

$$\begin{aligned} \phi^{-1}(\overline{C}) &= \{p \in \alpha X | \phi(p) \in \overline{B \cap X}\} \\ &= \{p \in \alpha X | p \in \mathrm{Cl}_{\alpha X}(B \cap X)\} \\ &= \mathrm{Cl}_{\alpha X}(B \cap X) = B. \end{aligned}$$

8.4 Remarks. The equivalence of (1) and (3) is due to Steiner [1968]. The equivalence of (1) and (2) is stated (without proof or attribution) in Šapiro [1974].

Frink [1964] asked whether or not all elements of $K(X)$ can be realized as $\omega_{\mathcal{C}}X$ for an appropriate normal base $\mathcal{C}$. This question is still unanswered, although many specific compactifications have been shown to be of this type (i.e., a Wallman-Frink compactification). We conclude this chapter with a reduction of this problem which has recently been effected. What has been shown is that if every compactification of the discrete spaces are Wallman-Frink, then all Hausdorff compactifications are Wallman-Frink. The first to prove this was Šapiro [1974]. Independently, Ünlü [1976] showed this for locally compact spaces. Steiner and Steiner [1976] removed this restriction. We follow the general outline of Šapiro although we simplify his one difficult proof by using a lemma of Ünlü as proved in Steiner and Steiner.

8.5 Lemma (Ünlü, Steiner-Steiner). Let $\alpha_1 X_1$ and $\alpha_2 X_2$ be compactifications of X_1 and X_2, respectively. If $F: \alpha_1 X_1 \to \alpha_2 X_2$ (onto) takes X_1 onto X_2 and is 1 - 1 on $\alpha_1 X_1 \setminus X_1$ and if $\alpha_1 X_1$ is a Wallman-Frink compactification of X_1 then $\alpha_2 X_2$ is a Wallman-Frink compactification of X_2.

Proof. We may assume $\alpha_1 X_1$ has a base ring $\mathcal{B}_1$ of closed sets satisfying condition (3) of 8.3. Because F is continuous and onto, and because $\alpha_2 X_2$ is compact, F is closed and $\mathcal{B}_2' = \{F(B_1) | B_1 \in \mathcal{B}_1\}$ is a base for the closed sets in $\alpha_2 X_2$. Let $\mathcal{B}_2$ be the family of all finite intersections of elements of $\mathcal{B}_2'$. Then $\mathcal{B}_2$ is a base ring of closed sets in $\alpha_2 X_2$. Now suppose $F(B_1),\dots,F(B_n) \in \mathcal{B}_2'$ and $G = F(B_1) \cap\dots\cap F(B_n) \neq \emptyset$. (In other words, G is a typical non-empty element of $\mathcal{B}_2$.) If $G \cap X_2 = \emptyset$ then $G \subset \alpha_2 X_2 \setminus X_2$ and so for $y \in G$ we have $y = F(b_1) =\dots= F(b_n)$ where $b_i \in B_i$, $i = 1,\dots,n$. No b_i can belong to X_1 since $F(X_1) = X_2$ and this would put y in X_2. Thus, $b_i \in \alpha_1 X_1 \setminus X_1$ for all i. By hypothesis, F is $1 - 1$ on $\alpha_1 X_1 \setminus X_1$. We conclude that $b_1 = b_2 =\dots= b_n$. Thus, $B_1 \cap\dots\cap B_n \neq \emptyset$. Now, $B_1 \cap\dots\cap B_n \in \mathcal{B}_1$ so that $(B_1 \cap\dots\cap B_n) \cap X_1 \neq \emptyset$ (Condition (3) of 8.3). But then

$$G \cap X_2 = [F(B_1) \cap\dots\cap F(B_n)] \cap X_2 \neq \emptyset,$$

a contradiction. We conclude that $\mathcal{B}_2$ does indeed satisfy Condition (3) of 8.3.

<u>8.6 Notation</u>. Let $\mathcal{W}_D$, $\mathcal{W}_C$, $\mathcal{W}$ denote the following statements:

$\mathcal{W}_D$: Every compactification of a discrete space is a Wallman-Frink compactification.

$\mathcal{W}_C$: Every compactification of a completely regular space with no isolated points is a Wallman-Frink compactification.

$\mathcal{W}$: Every compactification of a completely regular space is a Wallman-Frink compactification.

We wish to show that $\mathcal{W}_D$ implies $\mathcal{W}$. We first show $\mathcal{W}_D$ implies $\mathcal{W}_C$. We will do this using 8.5 and a construction which is originally due to Alexandroff-Urysohn [1929] and which was generalized to the form we use here by Engelking [1968].

<u>8.7 Construction</u>. Let X be a space without isolated points, let αX be a compactification of X, and let D be a discrete space with $|D| = |X|$. We construct a compactification, $\varepsilon_{\alpha X} D$, as follows: Let $f\colon D \to X$ be $1 - 1$ and onto, and let $\varepsilon_{\alpha X} D$ have as underlying set, $D \cup \alpha X$. Give it a topology by declaring points of D to be open and basic neighborhoods of points p in αX are of the form $V \cup [f^{-1}(V) \setminus f^{-1}(p)]$ where V is an original neighborhood of p (in αX).

Because X has no isolated points, it follows that D is dense in $\varepsilon_{\alpha X} D$. Because D is discrete, it follows that $\varepsilon_{\alpha X} D$ is Hausdorff. Because αX is compact, it follows that $\varepsilon_{\alpha X} D$ is compact.

8.8 Remark. Note that the above construction provides a result which could belong in Chapter 6 or 7: If αX has density character (6.35) k then it is a remainder of the discrete space of cardinality k.

8.9 Lemma. $\mathcal{W}_D$ implies $\mathcal{W}_C$.

Proof. If X has no isolated points, we may construct $\varepsilon_{\alpha X}D$ as above. By the hypothesis $\mathcal{W}_D$ it follows that $\varepsilon_{\alpha X}D$ is a Wallman-Frink compactification of D. By 8.5, αX is a Wallman-Frink compactification of X. We define $F\colon \varepsilon_{\alpha X}D \to \alpha X$ by

$$F(p) = \begin{cases} f(p) & \text{if } p \in D \\ p & \text{if } p \in \alpha X . \end{cases}$$

It is completely routine to verify that F is continuous and onto, that $F(D) = X$, and that F is $1-1$ on $\varepsilon_{\alpha X}D \setminus D$.

8.10 Lemma. Suppose $X = Y \cup Z$. If $\overline{Y}$ and $\overline{Z}$ are Wallman-Frink compactifications of Y and Z, respectively (where the closures are in αX), then αX is a Wallman-Frink compactification of X.

Proof. $\alpha X = \overline{Y} \cup \overline{Z}$. Let $W = \overline{Y} \cap \overline{Z}$. Let $\mathcal{A}$ and $\mathcal{B}$ be base rings in $\overline{Y}$ and $\overline{Z}$, respectively, which satisfy condition (2) of 8.3. Define

$$\mathcal{C} = \{A \cup B \mid A \in \mathcal{A},\ B \in \mathcal{B},\ \text{and } A \cap W \subset B \cap W\}.$$

It is routine to verify that $\mathcal{C}$ is a ring of closed sets in αX. We show that it is a base for the closed sets. Suppose $p \in \alpha X$, F is closed in αX, and $p \notin F$. There exists $A \in \mathcal{A}$ such that $p \notin A$ and $F \cap \overline{Y} \subset A$. Now $A \cap W$ is closed in αX and $p \notin (F \cap \overline{Z}) \cup (A \cap W)$. Thus, there exists $B \in \mathcal{B}$ such that $p \notin B$ and $(F \cap \overline{Z}) \cup (A \cap W) \subset B$. It is now apparent that $\mathcal{C}$ is a base for the closed sets in αX. If $C = A \cup B \in \mathcal{C}$ then $\overline{C \cap X} = \overline{(A \cup B) \cap X} = \overline{(A \cap X) \cup (B \cap X)} = (\overline{A \cap X}) \cup (\overline{B \cap X}) = A \cup B = C$. Thus by 8.3 (2), αX is a Wallman-Frink compactification of X.

8.11 Theorem. $\mathcal{W}_D$ implies $\mathcal{W}$.

Proof. Let D be the set of isolated points of X and let $C = X \setminus \overline{D}$ (closure in X). Then C has no isolated points and if αX is any compactification of X, it follows that αX is a compactification of $C \cup D$. D is discrete so $\overline{D}$ (closure in αX) is a Wallman-Frink compactification of D. Since C has no isolated points, $\overline{C}$ (closure in αX) is a Wallman-Frink compactification of C (8.9). By 8.9 we conclude that αX is a Wallman-Frink compactification of $C \cup D$. By

8.3 (2), αX has a base ring $\mathcal{A}$ of closed sets such that $\overline{A \cap (C \cup D)} = A$ for all $A \in \mathcal{A}$. But then $\overline{A \cap X} = \overline{A \cap (C \cup \overline{D})} = A$ for all $A \in \mathcal{A}$. We conclude that αX is a Wallman-Frink compactification of X.

<u>8.12 Major Problem #5</u>. The problem is whether or not $\mathcal{W}$ is true. We have seen that it is equivalent to whether or not $\mathcal{W}_D$ is true. (Trivially, $\mathcal{W}$ implies $\mathcal{W}_D$.) A naive approach to this problem might center around the following type of reasoning. We have seen that by taking $\mathcal{C}$ to be all zero sets in X we get $\omega_{\mathcal{C}} X \approx \beta X$. So for a specific $\alpha X \in K(X)$, why not take $\mathcal{C}$ to be the zero sets of the functions in C_α? Frink [1964] cites a particularly transparent example (which he attributes to R. M. Brooks) showing that this approach is doomed to failure. Let $X = N$ and let $\alpha X = \omega N$, the one-point compactification. Now any subset of N is the zero set of some function in C_ω. For $A \subset N$, define

$$f(p) = \begin{cases} 0, & p \in A \\ 1/p, & p \notin A. \end{cases}$$

Thus the family of zero sets of the functions in C_ω and the family of all zero sets in N are the same. If $\mathcal{C}$ is this family of zero sets, then $\omega_{\mathcal{C}} X \approx \beta X \not\approx \omega X$.

BIBLIOGRAPHY

Aarts, J.M.:

1968 Every metric compactification is a Wallman-type compactification, Proc. Internat. Sympos. on Topology and its Applications (Herceg-Novi, 1968), pp. 29-34. Savez Društava Mat. i Astronom., Belgrade, 1969. MR 42(1971), 1233.

1971 On the dimension of remainders in extensions of product spaces, Colloq. Math. 22(1971), 233-238. MR 43(1972), 1005.

Aarts, J.M. and van Ende Boas, P.:

1966 Continua as remainders in compact extensions, Math. Centrum Amsterdam Afd. Zuivere Wisk. 1966, ZW-002, 6 pp. MR 33(1967), 830-831.

1967 Continua as remainders in compact extensions, Nieuw Arch. Wisk. (3) 15 (1967), 34-37. MR 35(1968), 903.

Alexandroff, A.D.:

1942 On the extension of a Hausdorff space to an H-closed space, C.R. (Doklady) Acad. Sci. URSS (N.S.) 37(1942), 118-121. MR 5(1944), 45.

Alexandroff, P.S.:

1924 Über die Metrisation der im kleinen topologischen Räume, Math. Ann. 92(1924), 294-301.

1939 Bikompakte Erweiterungen topologischer Räume, Rec. Math (Mat. Sbornik) N.S. 5(47) (1939), 403-423. MR 1(1940), 318.

1947 On the concept of space in topology, Uspehi Matem. Nauk (N.S.) 2(1947), 5-57. MR 10(1949), 389.

1951 On the components of maximal bicompact extensions, Moskov. Gos. Univ. Učenye Zapiski 148, Matematika 4(1951), 216-218. MR 14(1953), 303.

1956 On two theorems of Yu. Smirnov in the theory of bicompact extensions, Fund. Math. 43(1956), 394-398. MR 18(1957), 813.

1960 Some results in the theory of topological spaces obtained within the last twenty-five years, Uspehi Mat. Nauk 15(1960), No. 2(92), 25-95. Translated as Russian Math. Surveys 15(1960), No. 2, 23-83. MR 22(1961), 1697.

1964 Some fundamental directions in general topology, Uspehi Mat. Nauk 19(1964), No. 6 (120), 3-46. MR 30(1965), 477.

Alexandroff, P.S. and Hopf, H.
1935 Topologie, I, Springer-Verlag, Berlin, 1935.

Alexandroff, P. and Ponomarev, V.:
1958 On bicompact extensions of topological spaces, Dokl. Akad. Nauk SSSR 121(1958), 575-578. MR 20(1959), 706.

1959 Compact extensions of topological spaces, Vestnik Moskov. Univ. Ser. Mat. Meh. Astr. Fiz. Him. 1959, No. 5, 93-108. MR 22(1961), 1698.

1962 On completely regular spaces and their bicompactifications, Vestnik Moskov. Univ. Ser. I Mat. Meh. 1962, No. 2, 37-43. MR 25(1963), 681.

Alexandroff, P.S. and Urysohn, P.:
1924 Zur theorie der topologischen Räume, Math. Ann. 92(1924), 258-266.

1929 Mémoire sur les espaces topologiques compacts, Verh. Nederl. Akad. Wetensch. Afd. Naturk. Sect. I, 14(1929), 1-96.

Alfsen, E.M. and Fenstad, J.E.:
1960 A note on completion and compactification, Math. Scand. 8(1960), 97-104. MR 23(1962), 674-675.

Allen, K.R.:
1975 Dendritic compactification, Pacific J. Math. 57(1975), 1-10.

Alò, R.A. and Sennott, L.I.:
1971 Extending linear space-valued functions, Math. Ann. 191(1971), 79-86. MR 43(1972), 1248-1249.

1972 Collectionwise normality and the extension of functions on product spaces, Fund. Math. 76(1972), 231-243. MR 48(1974), 205.

Alò, R.A. and Shapiro, H.L.:
1968 Normal bases and compactifications, Math. Ann. 175(1968), 337-340. MR 36(1968), 669.

1968 A note on compactifications and semi-normal bases, J. Austral. Math. Soc. 8(1968), 102-108. MR 37(1969), 647-648.

1969 Wallman compact and realcompact spaces, Contributions to Extension Theory of Topological Structures (Proc. Sympos., Berlin, 1967), 9-14. Deutsch. Verlag Wissensch., Berlin, 1969. MR 40(1970), 163.

1969 Z-realcompactifications and normal bases, J. Austral. Math. Soc. 9(1969), 489-495. MR 39(1970), 631-632.

1974 Normal topological spaces, Cambridge University Press, Cambridge, 1974, 306 pp.

Alò, R.A., Shapiro, H.L. and Weir, M.
1975 Realcompactness and Wallman realcompactification, Portugal. Math. 34(1975), 33-43.

Araya Muñoz, J.E.A.:
1963 Invariant measures on compactifications of the integers, Thesis, University of Washington, 1963.

Arhangel'skii, A. and Taimanov, A.:
1960 On a theorem of V. Ponomarev, Dokl. Akad. Nauk SSSR 135(1960), 247-248. Translated in Soviet Math. Dokl. 1(1961), 1242-1243. MR 23(1962), 547-548.

Arnautov, V.I.:
1962 Certain classes of completely regular spaces, Kišinev. Gos. Univ. Učen. Zap. 50(1962), 13-18. MR 36(1968), 187.

Atalla, R.E.:
1973 Regular matrices and P-sets in $\beta N \setminus N$, Proc. AMS 37(1973), 157-162. MR 48(1974), 525.

Aull, C.E.:
1972 Properties of side points of sequences, General Topology and Appl. 1(1971), 201-208. MR 44(1972), 866.

Baayen, P.C. and Paalman-de Miranda, A.B.:
1963 Disjoint open and closed sets in the complement of a discrete space in its Čech-Stone compactification, Math. Centrum Amsterdam Afd. Zuivere Wisk. 1963, ZW-008, 3 pp. MR 33(1967), 125.

Banaschewski, B.:
1955 Über nulldimensionale Räume, Math. Nachr. 13(1955), 129-140. MR 19(1958), 157.

1955 Über der Ultrafilterraum, Math. Nachr. 13(1955), 273-281. MR 17(1956), 179.

1956 Überlagerungen von Erweiterungs-räumen, Arch. Math. 7(1956), 107-115. MR 18(1957), 224.

1956 Local connectedness of extension spaces, Canad. J. Math. 8(1956), 395-398. MR 17(1956), 1229.

1959 On the Katetov and Stone-Čech extensions, Canad. Math. Bull. 2(1959), 1-4. MR 21(1960), 708.

1960 On homeomorphisms between extension spaces, Canad. J. Math. 12(1960), 252-262. MR 22(1961), 683.

1962 Normal systems of sets, Math. Nachr. 24(1962), 53-75. MR 30(1965), 478.

1963 On Wallman's method of compactification, Math. Nachr. 27(1963), 105-114. MR 28(1964), 666.

1964 Extensions of topological spaces, Canad. Math. Bull. 7(1964), 1-22. MR 28(1964), 876.

1974 A remark on extensions of Hausdorff spaces, General Topology and Appl. 4(1974), 283-284.

Banilower, H.:
1973 The Stone-Čech compactification of a basically disconnected space, Yokohama Math. J. 21(1973), 33-36. MR 48(1974), 871.

Behrend, F.A.:
1956 Note on the compactification of separated uniform spaces, Nederl. Akad. Wentensch. Proc. Ser. A 59(1956), 269-270. (= Indag. Math. 18). MR 17(1956), 1230.

1957 Uniformizability and compactifiability of topological spaces, Math. Z 67(1957), 203-210. MR 19(1958), 298-299.

Bellamy, D.P.:
1968 Topological Properties of Compactifications of a Half-open Interval, Thesis, Michigan State University, 1968.

1971 Aposyndesis in the remainder of Stone-Čech compactifications, Bull. Acad. Polon. Sci. Ser. Sci. Math. Astronom. Phys. 19(1971), 941-944. MR 46(1973), 1409.

Bellamy, D.P. and Rubin, L.R.:
1973 Indecomposable continua in Stone-Čech compactifications, Proc. AMS 39(1973), 427-432. MR 47(1974), 737.

Bentley, H.L.:
1972 Some Wallman compactifications determined by retracts, Proc. AMS 33(1972), 587-593. MR 45(1973), 488.

1972 Some Wallman compactifications of locally compact spaces, Fund. Math. 75(1972), 13-24. MR 46(1973), 767.

1972 Normal bases and compactifications, Proceedings of the University of Oklahoma Conference, 23-37. University of Oklahoma, Norman, Oklahoma, 1972.

Bentley, H.L. and Naimpally, S.A.:
1974 Wallman T_1-compactifications as epireflections, Gen. Topol. and its Appl. 4(1974), 29-41. MR 51(1976), 238.

Bentley, H.L. and Taylor, B.J.:
1975 Wallman rings, Pacific J. Math. 58(1975), 15-35.

Biles, C.M.:
1970 Wallman-type compactifications, Proc. AMS 25(1970), 363-368. MR 41(1971), 1400.

1970 Gelfand and Wallman-type compactifications, Pacific J. Math. 35(1970), 267-278. MR 43(1972), 1003.

Blair, R.L. and Hager, A.:
1974 Extensions of zero sets and of real-valued functions, Math. Z. 136(1974), 41-52.

1975 Notes on the Hewitt realcompactification of a product, Gen. Topol. and its Appl. 5(1975), 1-8.

Blakley, G.R., Gerlits, J. and Magill, K.D., Jr.
1971 A class of spaces with identical remainders, Studia Sci. Math. Hungar. 6(1971), 117-122. MR 50(1975), 2009-2010.

Blanksma, T.:
1967 Lattice characterizations of topologies and compactifications, Doctoral dissertation, University of Utrectht, 1968, 47 pp. MR 37(1969), 1092.

Blass, A.R.:
1973 The Rudin-Keisler ordering of P-points, Trans. AMS 179(1973), 145-166. MR 50(1975), 941.

Blatter, J.:
1975 Order compactifications of totally ordered topological spaces, J. Approximation Theory 13(1975), 56-65. MR 50(1975), 1529-1530.

Boboc, N. and Siretchi, Gh.
1964 Sur la compactification d'un espace topologique, Bull. Math. Soc. Sci. Math. Phys. R.P. Roumaine (N.S.) 5(53) (1964), 155-165. MR 32(1966), 81.

Brady, G.M.:
1975 Some results on (E, βE)-compactness, Studies in Topology, 81-92. Academic Press, New York, 1975. MR 51(1976), 239.

Brooks, R.M.:
1967 On Wallman compactifications, Fund. Math. 60(1967), 157-173. MR 35(1967), 182.

Brown, R.:
1974 On sequentially proper maps and a sequential compactification, J. London Math. Soc. (2) 7(1974), 515-522. MR 48(1974), 1659.

Brümmer, G.C.L.:
1972 Note on a compactification due to Nielsen and Sloyer, Math. Ann. 195(1972), 167. MR 45(1973), 204.

Burghelea, D.:
1962 Sur la compactification des espaces topologiques, Com. Acad. R.P. Române 12(1962), 667-670. MR 28(1964), 121.

Cain, G.L., Jr.:
1969 Compactification of mappings, Proc. AMS 23(1969), 298-303. MR 40(1970), 367.

1971 Extensions and compactifications of mappings, Math. Ann. 191(1971), 333-336. MR 44(1972), 617.

1972 Metrizable mapping compactifications, General Topology and Appl. 2(1972), 271-275. MR 46(1973), 1086.

Calder, A.:
1972 The cohomotopy groups of Stone-Čech increments, Indag. Math. 34(1972), 37-44. MR 46(1973), 456.

Carlson, D.H.:
1969 Critical points on rim-compact spaces, Pacific J. Math. 29(1969), 63-65. MR 39(1970), 1147.

Čech, E.:
1937 On bicompact spaces, Ann. of Math. 38(1937), 823-844.

1966 Topological Spaces, Publishing House of the Czechoslovak Academy of Sciences, Prague; Interscience Publishers, London-New York-Sidney, 1966. MR 35(1968), 426-427.

Čech, Eduard and Novák, Josef:
1948 On regular and combinatorial imbedding, Casopis Pěst Mat. Fys. 72(1947), 7-16. MR 9(1948), 98.

Chandler, R.E.:
1972 New compactifications from old, Amer. Math. Monthly 79(1972), 501-503. MR 45(1973), 1686-1687.

1972 An alternative construction of βX and υX, Proc. AMS 32(1972), 315-318. MR 45(1973), 204.

1976 Continua as remainders, revisited, Gen. Topol. and Appl., to appear.

Chandler, R.E. and Gellar, R.:
1973 The compactifications to which an element of $C^*(X)$ extends, Proc. AMS 38(1973), 637-639. MR 47(1974), 449.

Chou, C.
1969 Minimal sets and ergodic measures for $\beta N \setminus N$, Ill. J. Math. 13(1969), 777-788. MR 40(1970), 518.

Comfort, W.W.:
1963 An example in density character, Arch. Math. 14(1963), 422-423. MR 28(1964), 876.

1965 Retractions and other continuous maps from βX onto $\beta X \setminus X$, Trans. AMS 114(1965), 1-9. MR 32(1966), 513.

1967 A nonpseudocompact product space whose finite subproducts are pseudocompact, Math. Ann. 170(1967), 41-44. MR 35(1968), 182.

1968 A theorem of Stone-Čech type, and a theorem of Tychonoff type, without the axiom of choice; and their realcompact analogues, Fund. Math. 63(1968), 97-100. MR 38(1969), 922.

1969 Addendum to a paper of J. de Groot, Bull. Acad. Polon. Sci. Sér. Sci. Math. Astronom. Phys. 17(1969), 361-362. MR 41(1971), 1395.

Comfort, W.W. and Gordon, H.:
1964 Disjoint open subsets of $\beta X \setminus X$, Trans. AMS 111(1964), 513-520. MR 29(1965), 116.

Comfort, W.W. and Hager, A.W.:
1970 Estimates for the number of real-valued continuous functions, Trans. AMS 150(1970), 619-631. MR 41(1971), 1397.

1971 The projection mapping and other continuous functions on a product space, Math. Scand. 28(1971), 77-90. MR 47(1974), 735.

Comfort, W.W. and Negrepontis, S.:
1966 Extending continuous functions on X × Y to subsets of βX × βY, Fund. Math. 59(1966), 1-12. MR 34(1967), 132.

1968 Homeomorphs of three subspaces of βN∖N, Math. Z. 10(1968), 53-58. MR 38(1969), 502.

1974 The theory of ultrafilters, Springer-Verlag, New York, Heidelberg, 1974. x + 482 pp.

1975 Continuous pseudometrics, Marcel Dekker, Inc., New York, 1975. vi + 126 pp.

Conway, J.B.:
1966 Projections and retractions, Proc. AMS 17(1966), 843-847. MR 33(1967), 558.

Császár, Á.:
1962 Complétion et compactification d'espaces syntopogènes, General Topology and its Relations to Modern Analysis and Algebra (Proc. Sympos., Prague, 1961), 133-137. Academic Press, New York; Publ. House Czech. Acad. Sci., Prague, 1962. MR 33(1967), 327.

1964 Double compactification d'espaces syntopogènes, Ann. Univ. Sci. Budapest. Eötvös Sect. Math. 7(1964), 3-11. MR 30(1965), 653.

1968 Über die doppelte Kompaktifizierung gewisser topogener Räume, Ann. Univ. Sci. Budapest. Eötvös Sect. Math. 11(1968), 83-103. MR 39(1970), 1135-1136.

1971 Wallman-type compactifications and superextensions, Period. Math. Hungar. 1(1971), 55-80. MR 45(1973), 1405.

1972 Doppeltkompakte bitopologische Räume, Theory of Sets and Topology, 59-67. VEB Deutsch. Verlag Wissensch., Berlin, 1972. MR 49(1975), 1467.

1974 Function classes, compactifications, realcompactifications, Ann. Univ. Sci. Budapest. Eötvös Sect. Math. 17(1974), 139-156.

Császár, Á. and Mrowka, S.:
1959 Sur la compactification des espaces de proximité, Fund. Math. 46(1959), 195-207. MR 20(1959), 1188.

Cullen, Helen F.:
1968 Introduction to general topology, D. C. Heath and Co., Boston, 1968. MR 36(1968), 887.

Daguenet, M.:
1974 Propriété de Baire de βN muni d'une nouvelle topologie et application à la construction d'ultrafiltres, Séminaire Choquet, (1974/75), Initiation à l'analyse, Exp. No. 14, 3 pp.

1975 Rapport entre l'ensemble des ultrafiltres admettant un ultrefiltre donné pour image et l'ensemble des images de cet ultrafiltre, Comment. Math. Univ. Carolinae 16(1975), 99-113. MR 51(1976), 943-944.

Danzig, F.:
1969 Über die Äquivalenz von Erweiterungen, Contributions to Extension Theory of Topological Structures (Proc. Sympos., Berlin, 1967), 55-58. Deutsch. Verlag Wissensch., Berlin, 1969. MR 39(1970), 1365.

D'Aristotle, A.J.:
1971 Completely regular compactifications, Fund. Math. 71(1971), 139-145. MR 45(1973), 204-205.

Davis, G.:
1975 Automorphic compactifications and the fixed point lattice of a totally-ordered set, Bull. Austral. Math. Soc. 12(1975), 101-109. MR 51(1976), 241.

Day, M.M.:
1962 Normed linear spaces, Springer-Verlag, Berlin, 1962. MR 20(1959), 196.

Deák, E. and Hamburger, P.:
1973 Vollständig interne Charakterisierungen der T_2-kompaktifizierbaren Räume, Period. Math. Hungar. 4(1973), 125-145. MR 50(1975), 2007.

de Groot, J.:
1942 Topologische Studien. Compactificatie, Voorzetting van Afbeeldingen en Samenhang (Topological Studies. Compactification, Extension of mappings and Connectivity), Thesis, Univ. of Groningen (1942), 102 pp. (Dutch), MR 7(1946), 135-136.

de Groot, J. and Aarts, J.M.:
1969 Complete regularity as a separation axiom, Canad. J. Math. 21(1969), 96-105. MR 38(1969), 920.

de Groot, J. and McDowell, R.H.:
1959 Extensions of mappings on metric spaces, Fund. Math. 48(1959/60), 251-263. MR 23(1962), 239.

1967 Locally connected spaces and their compactifications, Ill. J. Math. 11(1967), 353-364. MR 35(1968), 1131.

de Groot, J., Jensen, G.A., and Verbeek, A.:
1968 Superextensions, Math. Centrum Amsterdam Afd. Zuivere Wisk. 1968 ZW-017, 33 pp. MR 40(1970), 1182.

de Groot, J., Hursch, J.L., Jr., and Jensen, G.A.:
1972 Local connectedness and other properties of GA compactifications, Nederl. Akad. Wetensch. Proc. Ser A 75 - Indag. Math. 34(1972), 11-18. MR 46(1973), 767-768.

Deprit, A.:
1956 Sur les M-compactifications d'Alexandroff, Acad. Roy. Belg. Bull. Cl. Sci. (5) 42(1956), 266-269. MR 17(1956), 1116.

Detourbet, G.:
1971 Espaces à fermetures, Esquisses mathematiques, No. 16, iv + 84 pp. Fac. Sci. Univ. Paris VII, Paris, 1971. MR 50(1975), 1147.

Dickman, R.F., Jr.:
1967 Compactness of mappings on products of locally connected generalized continua, Proc. AMS 18(1967), 1093-1094. MR 36(1968), 186.

1968 Some characterizations of the Freudenthal compactification of a semicompact space, Proc. AMS 19(1968), 631-633. MR 37(1969), 168.

1972 A necessary and sufficient condition for $\beta X \setminus X$ to be an indecomposable continuum, Proc. AMS 33(1972), 191-194. MR 45(1973), 798.

Dodziuk, J.:
1969 On measurable periodic functions (Polish), Wiadom. Math. 11(1969), 13-14. MR 40(1970), 639.

Duda, R.:
1964 On compactification of absolute retracts, Colloq. Math. 12(1964), 1-5. MR 29(1965), 983.

Dvališvili, B.P.:
1973 Complete regularity in terms of nets (Russian, English Summary), Sakharth, SSR Mech. Akad. Moambe 72(1973), 297-299. MR 49(1975), 1133.

Efimov, B.A.:
1969 Solution of certain problems on dyadic bicompacta, Dokl. Adad. Nauk SSSR 187(1969), 21-24. MR 40(1970), 1182.

1972 On the embedding of extremally disconnected spaces into bicompacta, General Topology and its Relations to Modern Analysis and Algebra, III, 103-107. Academia, Prague, 1972. MR 50(1975), 1152.

1975 The cardinality of extensions of dyadic spaces (Russian), Mat. Sb. (N.S.) 96(138) (1975), 614-632, 646-647. MR 51(1976), 1265-1266.

Engelking, R.:
1959 Sur la compactification des espaces metriques, Fund. Math. 48(1959/60), 321-324. MR 23(1962), 239.

1961 On the Freudenthal compactification, Bull. Acad. Polon. Sci. Sér. Sci. Math. Astronom. Phys. 9(1961), 379-383. MR 26(1963), 141.

1963 Topological spaces and their compactifications, Wiadom. Mat. (2) 6(1963), 135-172. MR 28(1964), 121.

1963 On certain compactifications of topological spaces, Prace Mat. 8(1963/64), 33-44. MR 32(1966), 290.

1964 Remarks on real-compact spaces, Fund. Math. 55(1964), 303-308. MR 31(1966), 723-724.

1968 Outline of General Topology, North Holland, Amsterdam; PWN, Warsaw; Interscience, New York, 1968. MR 36(1968), 887-888.

1968 On the double circumference of Alexandroff, Bull. Acad. Polon. Sci. Sér. Sci. Math. Astron. Phys. 16(1968), 629-634. MR 39(1970), 177.

1968 Outline of General Topology, North Holland Publishing Co., Amsterdam and PWN-Polish Scientific Publishers, Warsaw, 1968.

1972 Hausdorff's gaps and limits and compactifications, Theory of Sets and Topology (in honor of Felix Hausdorff, 1868-1942), pp. 89-93. VEB Deutsch. Verlag Wissensch., Berlin, 1972. MR 49(1975), 1135.

Engelking, R. and Mrówka, S.:
1958 On E-compact spaces, Bull. Acad. Polon. Sci. Sér. Sci. Math. Astr. Phys. 6(1958), 429-436. MR 20(1959), 581.

Engelking, R. and Skljarenko, E.G.:
1963 On compactifications allowing extensions of mappings, Fund. Math. 53(1963), 65-79. MR 27(1964), 998.

Evstigneev, V.G.:
1971 A new characterization of the Wallman extension and its minimal analogue for the general case of T-spaces (Russian) Vestnik Moskov. Univ. Ser. I Mat. Mech. 26(1971), No. 2, 34-41. MR 44(1972), 192.

Facini, G.:
1969 Gli spazi di filtri nello studio delle compattizzazioni e delle strutture uniformi di spazio precompatto (English summary), Rend. Ist. Mat. Univ. Trieste 1(1969), 138-149. MR 41(1971), 1135.

Fan, Ky and Gottesman, N.:
1952 On compactifications of Freudenthal and Wallman, Nederl. Akad. Wetensch. Proc. Ser. A 55 (=Indag. Math 14) (1952), 504-510. MR 14(1953), 669.

Fell, J.M.G.:
1962 A Hausdorff topology for the closed subsets of a locally compact non-Hausdorff space, Proc. AMS 13(1962), 472-476. MR 25(1963), 499.

Fine, N.J. and Gillman, L.:
1960 Extension of continuous functions in βN, Bull. AMS 66(1960), 376-381. MR 23(1962), 104.

1963 Remote points of βR, Proc. AMS 13(1962), 29-36. MR 26(1963), 141.

Firby, P.A.:
1970 Finiteness at infinity, Proc. Edinburgh Math. Soc. (2) 17(1970/71), 299-304. MR 46(1973), 450.

1972 Extensions of continuous functions, J. London Math. Soc. (2) 5(1972), 15-20. MR 45(1973), 1687.

1973 Lattices and compactifications, I, II, III, Proc. London Math. Soc. (3) 27(1973), 22-50, 51-60, 61-68. MR 48(1974), 524-525.

Flachsmeyer, J.:
1961 Zur Spektralentwicklung topologischer Räume, Math. Ann. 144(1961), 253-274. MR 26(1963), 141-142.

1966 Zur Theorie der H-abgeschlossenen Erweiterungen, Math. Z. 94(1966), 349-381. MR 35(1968), 1357.

1969 Über Erweiterungen mit nulldimensional gelegenem Adjunkt, Contributions to Extension Theory of Topological Structures (Proc. Sympos., Berlin, 1967), 63-72. Deutsch. Verlag Wissensch., Berlin, 1969. MR 40(1970), 365.

Fleischer, I.:
1971 Generalized compactifications for initial topologies, Rev. Roumaine Math. Pures Appl. 16(1971), 1185-1191. MR 45(1973), 1405.

Fomin, S.:
1940 Erweiterungen topologischer Räume, Rec. Math. [Mat. Sbornik] N.S. 8(50) (1940), 285-294. MR 2(1941), 320.

1943 Extensions of topological spaces, Ann. of Math (2) 44(1943), 471-480. MR 5(1944), 45.

1958 On the connection between proximity spaces and the bicompact extensions of completely regular spaces, Dokl. Akad. Nauk SSSR 121(1958), 236-238. MR 20(1959), 706.

Forge, A.B.:
1961 Dimension preserving compactifications allowing extensions of continuous functions, Duke Math. J. 28(1961), 625-627. MR 29(1965), 983.

Franklin, S.P.:
1975 On products of countably compact spaces, Proc. AMS 48(1975), 236-238. MR 51(1976), 1268.

Freudenthal, H.:
1942 Neuaufbau der Endentheorie, Ann. of Math (2) 43(1942), 261-279. MR 3(1943), 315.

1945 Über die Enden diskreter Räume und Gruppen, Comment. Math. Helv. 17(1945), 1-38. MR 6(1945), 277.

1951 Kompaktisierungen und Bikompaktisierungen, Neder. Akad. Wetensch. Proc. Ser. A. 54(1951), 184-192. MR 12(1951), 728. (= Indagationes Math. 13(1951).)

1952 Enden und Primenden, Fund. Math. 39(1952), 189-210. MR 14(1953), 893.

Frič, R.:
1972 Sequential envelope and subspaces of the Čech-Stone compactification, General Topology and its Relations to Modern Analysis and Algebra, III, 123-126. Academia, Prague, 1972. MR 50(1975), 787.

Frink, O.:
1964 Compactifications and semi-normal spaces, Amer. J. Math. 86(1964), 602-607. MR 29(1965), 770.

Frolik, Z.:
1959 Generalizations of compact and Lindelöf spaces, Czech. Math. J. 9(84) (1959), 172-217. MR 21(1960), 708.

1960 The topological product of two pseudocompact spaces, Czech. Math. J. 10(85) (1960), 339-349. MR 22(1961), 1209.

1961 On approximation and uniform approximation of spaces, Proc. Japan Acad. 37(1961), 530-532. MR 25(1963), 305.

1962 Locally G_δ-spaces, Czech. Math. J. 12(87) (1962), 346-355. MR 26(1963), 1052-1053.

1967 Sums of ultrafilters, Bull. AMS 73(1967), 87-91. MR 34(1967), 641.

1967 Non-homogeneity of $\beta P \setminus P$, Comment. Math. Univ. Carolinae 8(1967), 705-709. MR 42(1971), 189-190.

1967 Types of ultrafilters on countable sets, Gen. Topology and its Relations to Modern Analysis and Algebra II (Proc. Second Prague Top. Sympos., 1966), pp. 142-143. Academia, Prague, 1967.

1967 Homogeneity problems for extremally disconnected spaces, Comment. Math. Univ. Carolinae 8(1967), 757-763. MR 41(1971), 1691.

1968 Fixed points of maps of extremally disconnected spaces and complete Boolean algebras, Bull. Acad. Polon. Sci. Sér. Sci. Math. Astron. Phys. 16(1968), 269-275. MR 38(1969), 312.

1968 Fixed points of maps of βN, Bull. AMS 74(1968), 187-191. MR 36(1968), 1139-1140.

Gagrat, M.S. and Naimpally, S.A.:
1971 Proximity approach to extension problems, Fund. Math. 71(1971), 63-76. MR 45(1973), 486-487.

1973 Wallman compactifications and Wallman realcompactifications, J. Austral. Math. Soc. 15(1973), 417-427. MR 49(1975), 1136.

Garcia-Mayanez, A.:
1970 Compactifications and metrization of topological spaces, Bol. Sci. Mat. Mexicana (2) 15(1970), 52-57.

Gazik, R.J.:
1974 Regularity of Richardson's compactification, Canad. J. Math. 26(1974), 1289-1293. MR 50(1975), 784.

Gelfand, I. and Kolmogorff, A.:
1939 On rings of continuous functions on topological spaces, Dokl. Akad. Nauk SSSR 22(1939), 11-15.

Gelfand, I. and Šilov, G.:
1941 Uber verschiedene Methoden der Einfuhrung der Topologie in die Menge der Maximalen Idealen eines normierten Ringes, Rec. Math. (Mat. Sbornik) N.S. 9(51) (1941), 25-39. MR 3(1942), 52.

Gerlits, J.:
1974 On G_δ p-spaces, Topics in topology (Proc. Colloq., Keszthely, 1972), 341-346. Colloq. Math. Soc. János Bolyai, V.8, North-Holland, Amsterdam, 1974. MR 50(1975), 2010.

Gerolini, Annamaria:
1951 Compactification des espaces séparés, C.R. Acad. Sci. Paris 232(1951), 1056-1058. MR 12(1951), 628.

Ghisa, D,:
1971 On the theorem of compactificability of topological spaces, An. Univ. Trinisoara Ser. Sti. Mat. 9(1971), 161-163. MR 48(1974), 525.

Gillman, L.:
1961 A note on F-spaces, Arch. Math. 12(1961), 67-68. MR 23(1962), 407.

1967 The space βN and the continuum hypothesis, General Topology and its Relations to Modern Analysis and Algebra, II (Proc. Second Prague Topological Sympos., 1966), pp. 144-146. Academia, Prague, 1967. MR 38(1969), 310.

Gillman, L. and Henriksen, M.:
1956 Rings of continuous functions in which every finitely generated ideal is principal, Trans. AMS 82(1956), 366-391. MR 18(1957), 9.

Gillman, L. and Jerison, M.:
1959 Stone-Čech compactification of a product, Arch. Math. 10(1959), 443-446. MR 22(1961), 39.

1960 Rings of Continuous Functions, D. Van Nostrand Co., Inc., Princeton, 1960. MR 22(1961), 1190-1191.

Ginsburg, J.:
1976 On the Stone-Čech compactification of the space of closed sets, Trans. AMS 215(1976), 301-311.

Glicksberg, I.:
1959 Stone-Čech compactifications of products, Trans. AMS 90(1959), 369-382. MR 21(1960), 815-816.

Gordon, H.:
1970 Compactifications defined by means of generalized ultrafilters, Ann. Mat. Pura Appl. (4) 86(1970), 15-23. MR 42(1971), 950.

Gould, G.G.:
1964 A Stone-Čech-Alexandroff-type compactification and its application to measure theory, Proc. London Math. Soc. (3) 4(1964), 221-244. MR 30(1965), 90.

Grinblat, L.Š.:
1976 Compactifications of spaces of functions and integration of functionals, Trans. AMS 217(1976), 195-223.

Guthrie, J.A.:
1976 Some discrete subspaces of βm, Rocky Mountain J. Math. 6(1976), 265-267.

Hadžrivanov, N.:
1967 The compactification of proximity spaces (Russian), General Topology and its Relations to Modern Analysis and Algebra, II (Proc. Second Prague Topological Sympos., 1966), 164-170. Academia, Prague, 1967. MR 38(1969), 921.

Hager, A.:
1966 Some remarks on the tensor product of function rings, Math. Z. 92(1966), 210-244. MR 33(1967), 326.

1973 Compactification and completion as absolute closure, Proc. AMS 40(1973), 635-638. MR 48(1974), 1662.

Hahn, H.:
1914 Mengentheoretische Characterisierung der Stetigen Kurven, Sitzungsberichte, Akad. der Wissenschaften 123(1914), 2433.

Hajek, D.W.:
1974 A note on Wallman extendible functions, Proc. AMS 44(1974), 505-506. MR 49(1975), 1799.

1974 Functions with continuous Wallman extensions, Czech. Math. J. 24(99) (1974), 40-43. MR 51(1976), 239.

Hajnal, A. and Juhász, I.:
1967 Discrete subspaces of topological spaces, Nederl. Akad. Wetensch. Proc. Ser. A 70 = Indag. Math. 29(1967), 343-356. MR 37(1969), 882.

Hamburg, I.:
1971 Observations on the compactification of S. Leader and Ju. M. Smirnov (Roumanian, English Summary), Stud. Cerc. Mat. 23(1971), 1353-1360. MR 48(1974), 1662.

Hamburger, P.:
1971 On Wallman-type, regular Wallman-type, and z-compactifications, Period. Math. Hungar. 1(1971), 303-309. MR 45(1973), 488.

1972 On k-compactifications and realcompactifications, Acta Math. Acad. Sci. Hungar. 23(1972), 255-262. MR 47(1974), 181.

Hardy, J.P.L., Morris, S.A., and Thompson, H.B.:
1976 Applications of the Stone-Čech compactification to free topological groups, Proc. AMS 55(1976), 160-164.

Hardy, K. and Woods, R.G.:
1972 On c-realcompact spaces and locally bounded normal functions, Pacific J. Math. 43(1972), 647-656. MR 51(1976), 240.

Harris, D.:
1971 The Wallman compactification as a functor, General Topology and Appl. 1(1971), 273-281. MR 45(1973), 205.

1971 Katetov extension as a functor, Math. Ann. 193(1971), 171-175. MR 45(1973), 1405.

1971 Structures in topology, Memoirs of the American Mathematical Society, No. 115, Providence, R.I., 1971. vi + 96 pp. MR 45(1973), 1098-1099.

1972 The Wallman compactification is an epireflection, Proc. AMS 31(1972), 265-267. MR 44(1972), 1080.

1974 Closed images of the Wallman compactification, Proc. AMS 42(1974), 312-319. MR 49(1975), 1466.

1974 Semirings and T_1 compactifications, I, Trans. AMS 188(1974), 241-258. MR 51(1976), 239.

Hayashi, E.:
1958 One point expansion of topological spaces, Proc. Japan Acad. 34(1958), 73-75. MR 21(1960), 426.

Hechler, S.H.:
1971 Classifying almost-disjoint families with applications to $\beta N \setminus N$, Israel J. Math. 10(1971), 413-432. MR 46(1973), 266-267.

1972 Short complete nested sequences in $\beta N \setminus N$ and small maximal almost-disjoint families, Gen. Top. and Appl. 2(1972), 139-149. MR 46(1973), 1214.

Heider, L.J.:
1956 Directed limits on rings of continuous functions, Duke Math. J. 23(1956), 293-296. MR 17(1956), 990.

1959 Compactifications of dimension zero, Proc. AMS 10(1959), 377-384. MR 21(1960), 708.

Henderson, D.W.:
1968 D-dimension. II. Separable spaces and compactifications, Pacific J. Math. 26(1968), 109-113. MR 39(1970), 876.

Henriksen, M.:
1957 On minimal completely regular spaces associated with a given ring of continuous functions, Mich. Math. J. 4(1957), 61-64. MR 18(1957), 916-917.

Henriksen, M. and Isbell, J.R.:
1957 On the Stone-Čech compactification of a product of two spaces, Bull. AMS 63(1957), 145-146.

1957 Some properties of compactifications, Duke Math. J. 25(1957), 83-105. MR 20(1959), 447.

1957 Local connectedness in the Stone-Čech compactification, Ill. J. Math. 1(1957), 574-582. MR 20(1959), 446-447.

Herrlich, H.:
1967 Fortsetzbarkeit stetiger Abbildungen und Kompaktheitsgrad topologischer Räume, Math. Z. 96(1967), 64-72. MR 34(1967), 1535.

1968 Topologische Reflexionen, und Coreflexionen, Lecture Notes in Mathematics, Springer-Verlag, Berlin-New York, 1968. MR 41(1971), 178.

Herrmann, R.A.:
1975 Nonstandard topological extensions, Bull. Austral. Math. Soc. 13(1975), 269-290.

Hewitt, E.:
1946 A remark on density characters, Bull. AMS 52(1946), 641-643.

1948 Rings of real-valued continuous functions I, Trans. AMS 64(1948), 54-99. MR 10(1949), 126-127.

1949 A note on extensions of continuous real functions, Anais Acad. Brasil. Ci. 21(1949), 175-179. MR 11(1950), 194.

Higman, G.:
1948 The compacting of topological spaces, Quart. J. Math., Oxford Ser. 19(1948), 27-32. MR 9(1948), 455.

Hindman, N.:
1969 On the existence of c-points in $\beta N \setminus N$, Proc. AMS 21(1969), 277-280. MR 39(1970), 177.

1973 Preimages of points under the natural map from $\beta(N \times N)$ to $\beta N \times \beta N$, Proc. AMS 37(1973), 603-608. MR 50(1975), 1530.

Holm, P.:
1962 C-completion and quasi-compactification, Avh. Norske Vid.-Akad. Oslo I (N.S.) No. 3(1962), MR 25(1963), 585.

Horelick, B.:
1969 Grouplike extensions and similar algebras, Math. Systems Theory 3 (1969), 139-145. MR 40(1970), 651.

Hopf, H.:
1944 Enden offener Räume and unendliche diskontinuierliche Gruppen, Comment. Math. Helv. 16(1944), 81-100. MR 5(1944), 272-273.

Hoshina, T.:
1973 On unions of countably-compactifiable spaces, Sci. Rep. Tokyo Kyoiku Daigaku Sect. A 12(1973), 92-94. MR 50(1975), 2009.

Hsieh, D.K.:
1974 The representation of the N-compactification by the λ-compactification, Math. Japan 19(1974), 57-61. MR 50(1975), 2008.

Hu, Sze-tsen:
1964 Elements of General Topology, Holden-Day, San Francisco, 1964. MR 31(1966), 298-299.

Hung, H.H. and Negrepontis, S.:
1974 Spaces homeomorphic to $(2^{\alpha})_{\alpha}$. II, Trans. AMS 188(1974), 1-30. MR 51(1976), 938.

Hunsacker, W.N. and Naimpally, S.A.:
1974 C-embedding of Hausdorff spaces in certain extensions, Topology Conference VPI, 1972. Lecture Notes in Math., V. 375, Springer, Berlin, 1974.

1974 Hausdorff compactifications as epireflections, Canad. Math. Bull. 17(1974/75), 675-677.

Hursch, J.L., Jr.:
1971 The local connectedness of GA compactifications generated by all closed connected sets, Nederl. Akad. Wetensch. Proc. Ser. A 74 = Indag. Math. 33(1971), 411-417. MR 46(1973), 1728.

Hušek, M.:
1969 Categorical connections between generalized proximity spaces and compactifications, Contributions to Extension Theory of Topological Structures (Proc. Sympos., Berlin, 1967), 127-132. Deutsch. Verlag Wissensch., Berlin, 1969. MR 40(1970), 362.

1972 Products as reflections, Comment. Math. Univ. Carolinae 13(1972), 783-800. MR 47(1974), 1672-1673.

1974 Hewitt realcompactifications of products, Topics in Topology (Proc. Colloq., Keszthely, 1972), Colloq. Math. Soc. János Bolyai, Vol. 8, pp. 427-435, North Holland, Amsterdam, 1974.

Iliadis, S.:
1963 Absolutes of Hausdorff spaces, Dokl. Akad. Nauk SSSR 149(1963), 22-25. MR 28(1964), 122-123.

1963 Some properties of absolutes, Dokl. Akad. Nauk SSSR 152(1963), 798-800. MR 27(1964), 808.

Iliadis, S. and Fomin, S.:
1966 The method of concentric systems in the theory of topological spaces, Uspehi Mat. Nauk 21(1966), No. 4(130), 47-76. MR 34(1967), 641-642.

Inagaki, T. and Sugawara, M.:
1952 Compactification of topological spaces, Math. J. Okayama Univ. 2(1952), 85-97. MR 14(1953), 489.

Inasaridze, H.N.:
1966 A generalization of perfect mappings, Dokl. Akad. Nauk SSSR 168(1966), 266-268. Translation: Soviet Math. Dokl. 7(1966), 620-622. MR 33(1967), 830.

1966 On extensions and growths of finite order for completely regular spaces, Dokl. Akad. Nauk SSSR 166(1966), 1043-1045. Translated: Soviet Math. Dokl. 7(1966), 229-231. MR 33(1967), 826.

Isbell, J.R.:
1955 Zero-dimensional spaces, Tohoku Math. J. (2) 7(1955), 1-8. MR 19(1958), 156-157.

1964 Uniform Spaces, Mathematical Surveys, No. 12, AMS, Providence, R.I., 1964. MR 30(1965), 113-114.

1965 Spaces without large projective subspaces, Math. Scand. 17(1965), 89-105. MR 33(1967), 827.

1968 Correction to "Spaces without large projective subspaces", Math. Scand. 22(1968), 310. MR 41(1971), 178.

1971 A corrected correction, Fund. Math. 71(1971), 185. MR 45(1973), 201.

Isiwata, T.:
1957 A generalization of Rudin's theorem for the homogeneity problem, Sci. Rep. Tokyo Kyoiku Daigaku. Sect A 5(1957), 300-303. MR 20(1959), 324.

1957 On subspaces of Čech-compactification space, Sci. Rep. Tokyo Kyoiku Daigaku. Sec. A. 5(1957), 304-309. MR 20(1959), 324.

1958 Normality of the product space of a countably compact space with its any compactification, Sci. Rep. Tokyo Kyoiku Daigaku Sect. A 6(1958), 181-184. MR 21(1960), 562.

1958 On Stonian spaces, Sci. Rep. Tokyo Kyoiku Daigaku Sect. A 6(1958), 147-176. MR 21(1960), 708.

1959 Some properties of F-spaces, Proc. Japan Acad. 35(1959), 71-76. MR 21(1960), 709.

1959 On the duality concerning Stone-Čech compactifications, Sûgaku 11(1959/60), 226-228. MR 25(1963), 118.

1960 Characterizations of spaces with dual spaces, Proc. Japan Acad. 36(1960), 200-204. MR 22(1961), 2152.

1963 Normality and perfect mappings, Proc. Japan Acad. 39(1963), 95-97. MR 27(1964),153.

1967 Mappings and spaces, Pacific J. Math. 20(1967), 455-480. Correction, ibid. 23(1967), 630-631. MR 36(1968), 436-437.

1969 Z-mappings and C*-embeddings, Proc. Japan Acad. 45(1969), 889-893. MR 42(1971), 191.

1974 Some properties of the remainder of Stone-Čech compactifications, Fund. Math. 83(1974), 129-142. MR 50(1975), 1530.

1974 On closed countably-compactifications, Gen. Topol. and its Appl. 4(1974), 143-167. MR 49(1975), 1799.

Ivanov, A.A.:

1966 Regular extensions of topological spaces, Vesci Akad. Navuk BSSR Ser. Fiz.-Mat. Navuk 1966, No. 1, 28-35. MR 35(1968), 181.

1959 Contiguity relations on topological spaces, Dokl. Akad. Nauk SSSR 128(1959), 33-36. MR 21(1960), 1393.

1973 Extension structures (Russian), Zap. Naučn. Sem. Leningrad. Otdel. Mat. Inst. Steklov. (LOMI) 36(1973), 126-133.

Ivanova, V.M.:

1960 Spaces of closed subsets of compact extensions, Mat. Sb. (N.S.) 50(92) (1960), 91-100. MR 22(1961), 505-506.

Ivanova, V.M. and Ivanov, A.A.:

1959 Contiguity spaces and bicompact extensions of topological spaces, Izv. Akad. Nauk SSSR Ser. Mat. 23(1959), 613-634. MR 22(1961), 167.

1959 Contiguity spaces and bicompact extensions of topological spaces, Dokl. Akad. Nauk SSSR 127(1959), 20-22. MR 21(1960), 1107-1108.

1972 Continuous mappings of extensions of a topological space, General Topology and its Relations to Modern Analysis and Algebra, Academia, Prague, 1972, pp. 209-214. MR 51(1976), 944.

Jayachandran, M. and Rajagopalan, M.:

1975 Scattered compactification for NU{P}, Pacific J. Math. 61(1975), 161-171.

Jech, T.J.:

1973 The Axiom of Choice, North-Holland Publishing Co., Amsterdam, 1973.

Jerison, M., Siegel, J., and Weingram, S.:

1969 Distinctive properties of Stone-Čech compactifications, Topology 8(1969), 195-201. MR 39(1970), 837.

Johnson, D.R.:

1971 The Construction of Elementary Extensions and the Stone-Čech Compactification, Thesis, Yale University, 1971.

Jongmans, F. and Moors, R.:
1961 Extensions et compactifications moniques d'un espace topologique, Bull. Soc. Roy. Sci. Liège 30(1961), 320-333. MR 23(1962), 546.

Juhász, I.:
1969 On the character of points in βN_m, Contributions to Extension Theory of Topological Structures (Proc. Sympos. Berlin, 1967), pp. 138-139. Deutscher Verlag Wissensch., Berlin, 1969.

Jung, C.F.K.:
1973 Locally compact spaces whose Alexandroff one-point compactifications are perfect, Colloq. Math. 27(1973), 247-249. MR 48(1974), 870.

Junghenn, H.D.:
1975 Almost periodic compactifications of transformation semigroups, Pacific Math. J. 57(1975), 207-216. MR 51(1976), 1609.

Kannan, V. and Thrivikraman, T.:
1975 Lattices of Hausdorff compactifications of a locally compact space, Pacific J. Math 57(1975), 441-444.

Katětov, M.:
1940 Über H-abgeschlossene und bikompakte Räume, Časopis Pěst. Mat. Fys. 69(1940), 36-49. MR 1(1940), 317-318.

1947 On the equivalence of certain types of extensions of topological spaces, Časopis Pěst. Mat. Fys. 72(1947), 101-106. MR 9(1948), 522.

1947 On H-closed extensions of topological spaces, Časopis Pěst. Mat. Fys. 72(1947), 17-32. MR 9(1948), 153.

1950 A theorem on Lebesgue dimension, Časopis Pěst. Mat. Fys. 75(1950), 79-87. MR 12(1951), 119.

1967 A theorem on mappings, Comment. Math. Univ. Carolinae 8(1967), 431-433. MR 37(1969), 887.

Kaufman, R.:
1967 Ordered sets and compact spaces, Coll. Math. 17(1967), 35-39. MR 35(1968), 662.

Keesling, J.:
1969 Compactification and the continuum hypothesis, Fund. Math. 66(1969/70), 53-54. MR 41(1971), 181.

1969 Open and closed mappings and compactifications, Fund. Math. 65(1969), 73-81. MR 40(1970), 365.

1971 Proper mappings and the minimum dimension of a compactification of a space, Proc. AMS 30(1971), 593-598. MR 44(1972), 1081-1082.

Kelley, J.L.:
1950 The Tychonoff product theorem implies the axiom of choice, Fund. Math. 37(1950), 75-76. MR 12(1951), 626.

1955 General Topology, D. Van Nostrand Co., Inc., New York, 1955. Reprinted by Springer-Verlag, Berlin, 1976. MR 16(1955), 1136-1138.

Kent, D.C., Richardson, G.D., and Gazik, R.J.:
1975 T-regular-closed convergence spaces, Proc. AMS 51(1975), 461-468. MR 51(1976), 1600.

Kim, J.:
1972 Sequentially complete spaces, J. Korean Math. Soc. 9(1972), 39-43. MR 46(1973), 451.

Kim, S.K.:
1969 Stone-Čech compactifications of infinite discrete spaces, J. Korean Math. Soc. 6(1969), 37-40. MR 48(1974), 208.

Kost, F.:
1971 Wallman-type compactifications and products, Proc. AMS 29(1971), 607-612. MR 43(1972), 1249.

1971 Finite products of Wallman spaces, Duke Math. J. 38(1971), 545-549. MR 43(1972), 1458.

1973 α-point compactifications, Portugal. Math. 32(1973), 132-137. MR 48(1974), 208.

Knaster, B.:
1952 Un théorème sur la compactification, Ann. Soc. Polon. Math. 25(1952), 252-267. MR 15(1954), 51.

Krivoručko, A.I.:
1974 The perfect extensions of topological spaces, Mat. Zametki 15(1974), 509-513, English Translation: Math. Notes 15(1974), 296-298. MR 50(1975), 1153.

Kroonenberg, N.S.:
1971 A topological compact Hausdorff space with countably many isolated points in which sets of isolated points cannot be left out, Bull. Acad. Polon. Sci. Sér. Sci. Math. Astronom. Phys. 19(1971), 501-503. MR 46(1973), 451.

Kulpa, W.:
1974 Remark on product of proximities, Topics in Topology, pp. 455-458. Colloq. Math. Soc. János Bolyai, V. 8, North-Holland, Amsterdam, 1974. MR 50(1975), 1153.

Kuznecova, T.A.:
1973 Continuous mappings and bicompact extensions of topological spaces, Vestnik Moskov. Univ. Ser. I Mat. Meh. 28(1973), 48-53. Translation: Moscow Univ. Math. Bull. 28(1973), 40-44. MR 49(1975), 258.

Lašnev, N.S.:
1969 Perfect irreducible mappings of completely regular spaces, Dokl. Akad. Nauk SSSR 188(1969), 282-285. Translation: Soviet Math. Dokl. 10(1969), 1111-1114. MR 40(1970), 907.

Lavallee, L.D.:
1965 The one-point countable compactification of curve spaces and arc spaces, Port. Math. 24(1965), 105-114. MR 33(1967), 558.

Leader, S.:
1969 Extensions based on proximity and boundedness, Math. Z. 108(1969), 137-144. MR 39(1970), 175-176.

Lelek, A.:
1962 On compactifications of some subsets of Euclidean spaces, Colloq. Math. 9(1962), 79-83. MR 24(1962), 546.

1965 On the dimensionality of remainders in compactifications, Dokl. Akad. Nauk SSSR 160(1965), 534-537. Translation: Soviet Math. Dokl. 6(1965), 136-140. MR 32(1966), 783-784.

Levine, J. and Dalton, R.E.:
1962 Minimum periods, modulo p, of first-order Bell exponential integers, Math. Comp. 80(1962), 416-423.

Levy, R. and McDowell, R.H.:
1975 Dense subsets of βX, Proc. AMS 50(1975), 426-430.

Liu, C.T. and Strecker, G.E.:
1972 Concerning almost realcompactifications, Czechoslovak Math. J. 22(97), (1972), 181-190. MR 46(1973), 451.

Liu, Chen-tung:
1968 Absolutely closed spaces, Trans. AMS 130(1968), 86-104. MR 36(1968), 433.

1969 The α-closure αX of a topological space X, Proc. AMS 22(1969), 620-624. MR 39(1970), 1140-1141.

Loeb, P.A.:
1967 A minimal compactification for extending continuous functions, Proc. AMS 18(1967), 282-283. MR 35(1968), 1357.

1969 Compactifications of Hausdorff spaces, Proc. AMS 22(1969), 627-634. MR 39(1970), 1141.

Lorch, E.R.:
1963 Compactification, Baire functions, and Daniell integration, Acta Sci. Math. (Szeged) 24(1963), 204-218. MR 29(1965), 116-117.

Louveau, A.:
1972 Ultrafilter sur N, et dérivation séquentielle, Bull. Sci. Math. (2) 96(1972), 353-382. MR 50(1975), 161.

1972 Dérivation séquentielle dans βN, C.R. Acad. Sci. Paris Sér. A-B 275(1972), A541-A544. MR 49(1975), 1466.

1973 Caractérisation des sous-espaces compacts de βN, Bull. Sci. Math. (2) 97(1973), 259-263. MR 50(1975), 787.

Lozier, F.W.:
1972 A compactification of locally compact spaces, Proc. AMS 31(1972), 577-579. MR 44(1972), 616.

Lubben, R.G.:
1941 Concerning the decomposition and amalgamation of points, upper semi-continuous collections, and topological extensions, Trans. AMS 49(1941), 410-466. MR 3(1942), 136.

Lukeš, J.:
1969 On the topological extensions, Comment. Math. Univ. Carolinae 10(1969), 407-420. MR 41(1971), 482.

Mack, J., Rayburn, M., and Woods, R.G.:
1972 Local topological properties and one point extensions, Canad. J. Math. 24(1972), 338-348. MR 45(1973), 798.

1974 Lattices of topological extensions, Trans. AMS 189(1974), 163-174. MR 50(1975), 444.

MacLane, S.:
1971 Categories for the Working Mathematician, Springer-Verlag, Berlin, 1971.

Magill, K.D., Jr.:
1965 N-point compactifications, Amer. Math. Monthly 72(1965), 1075-1081. MR 32(1966), 513.

1966 Countable compactifications, Canad. J. Math. 18(1966), 616-620. MR 33(1967), 1113.

1966 A note on compactifications, Math. Z. 94(1966), 322-325. MR 34(1967), 642.

1968 The lattice of compactifications of a locally compact space, Proc. London Math. Soc. (3) 18(1968), 231-244. MR 37(1969), 884.

1970 More on remainders of spaces in compactifications, Bull. Acad. Polon. Sci. Sér. Sci. Math. Astronon. Phys. 18(1970), 449-451. MR 42(1971), 677.

1970 Structure spaces of semigroups of continuous functions, Trans. AMS 149(1970), 595-600. MR 42(1971), 193.

1971 Homomorphisms and isomorphisms of semigroups of continuous self maps, General Topology and its Relation to Modern Analysis and Algebra, III (Proc. Conf., Kanpur, 1968), pp. 175-180. Academia, Prague, 1971. MR 43(1972), 741.

1971 K-Structure spaces of semigroups generated by idempotents, J. London Math. Soc. (2) 3(1971), 321-325. MR 43(1972), 1006.

1971 On the remainders of certain metric spaces, Trans. AMS 160(1971), 411-417. MR 44(1972), 616.

Makai, E., Jr.:
1972 Compactifications and a dual of compact spaces, Studia Sci. Math. Hungar. 7(1972), 199-200. MR 50(1975), 2009.

Malyhin, V.I.:
1972 Countable spaces having no bicompactifications of countable tightness, Dokl. Akad. Nauk SSSR 206(1972), 1293-1296. Translation: Soviet Math. Dokl. 13(1972), 1407-1411. MR 47(1974), 1672.

1975 Sequential bicompacta: Čech-Stone extensions and π-points, Vestnik Moskov. Univ. Ser. I Mat. Meh. 30(1975), 23-29. English Translation: Moskow Univ. Math. Bull. 30(1975), 18-23. MR 51(1976), 1604-1605.

Malyhin, V.I. and Sapirovskii, B.E.:
1973 Martin's axiom, and properties of topological spaces, Dokl. Akad. Nauk SSSR 213(1973), 532-535. English Translation: Soviet Math. Dokl. 14(1973), 1746-1751. MR 49(1975), 1466.

Mandelker, M.:
1969 Round z-filters and round subsets of βX, Israel J. Math. 7(1969), 1-8. MR 39(1970), 1141.

Manes, E.G.:
1974 Compact Hausdorff objects, Gen. Topol. and its Appl. 4(1974), 341-360. MR 51(1976), 581.

Marczewski, E.:
1947 Separabilité et multiplication cartésienne des espaces topologiques, Fund. Math. 34(1947), 127-143.

Mardešic, S.:
1960 On covering dimension and inverse limits of compact spaces, Illinois J. Math. 4(1960), 278-291. MR 22(1961), 1209.

Mattson, D.A.:
1970 Extension of proximity functions, Proc. AMS 26(1970), 347-351. MR 41(1971), 1699.

Mazurkiewicz, S.:
1920 Sur les lignes de Jordan, Fund. Math. 1(1920), 166-209.

1945 Recherches sur la théorie des bouts premiers, Fund. Math. 33(1945), 177-228. MR 8(1947), 47.

McArthur, W.:
1970 Hewitt realcompactifications of products, Canad. J. Math 22(1970), 645-656. MR 42(1971), 190.

McCallion, T.:
1972 Compactifications of ordered topological spaces, Proc. Cambridge Philos. Soc. 71(1972), 463-473. MR 45(1973), 490.

McCartney, J.R.:
1970 Maximal zero-dimensional compactifications, Proc. Cambridge Philos. Soc. 68(1970), 653-661. MR 42(1971), 950-951.

1971 Maximum countable compactifications of locally compact spaces, Proc. London Math. Soc. (3) 22(1971), 369-384. MR 44(1972), 868.

McDowell, R.H.:
1958 Extension of functions from dense subspaces, Duke Math. J. 25(1958), 297-304. MR 20(1959), 706.

Mednikov, L.E.:
1975 Imbedding of compact Hausdorff spaces in the Tychonoff cube and extending mappings from subsets of a product, Dokl. Akad. Nauk SSSR 222(1975), 1287-1290. Translated: Soviet Math. Dokl. 16(1975), 766-771.

Misonou, Y. and Takeda, Z.:
1952 On the compactification of topological spaces, Kodai Math. Sem. Rep. 1952, 17-18. MR 14(1953), 303.

Misra, A.K.:
1973 Some regular Wallman βX, Nederl. Akad. Wetensch. Proc. Ser. A 76 = Indag. Math. 35(1973), 237-242. MR 48(1974), 525.

Moors, R.:
1964 Extensions d'un espace topologique associées à une famille de tamis; compactifications d'un espace topologique, Bull. Soc. Roy. Sci. Liège 33(1964), 59-81. MR 29(1965), 770.

1966 Compactification des espaces topologiques, Bull. Soc. Roy. Sci. Liège 35(1966), 40-56. MR 34(1967), 132.

1968 Compactification d'espaces topologiques, Mém. Soc. Roy. Sci. Liège Coll. in-8° (5) 16(1968), No. 3, 73 pp. MR 38(1969), 129.

Morita, K.:
1951 On the simple extension of a space with respect to a uniformity, I, II, III, IV, Proc. Japan Acad. 27(1951), 65-72, 130-137, 166-171, 632-636. MR 14(1953), 68-69, 571.

1952 On bicompactifications of semibicompact spaces, Sci. Rep. Tokyo Bunrika Daigaku, Sec. A 4(1952), 222-229. MR 14(1953), 571.

1973 Countably compactifiable spaces, Sci. Rep. Tokyo Kyoiku Daigaku, Sect. A 12(1973), 7-15.

Mrówka, S.:
1956 Remark on P. Alexandroff's work "On two theorems of Yu. Smirnov", Fund. Math. 43(1956), 399-400. MR 18(1957), 813.

1959 On the potency of subsets of βN, Colloq. Math. 7(1959), 23-25. MR 22(1961), 843.

1970 Continuous functions on countable subspaces, Port. Math. 29(1970), 177-180. MR 45(1973), 491.

1971 Some consequences of Archangelskii's theorem, Bull. Acad. Polon. Sci. Sér. Sci. Math. Astronom. Phys. 19(1971), 373-376. MR 46(1973), 448.

1973 β-like compactifications, Acta Math. Acad. Sci. Hungar. 24(1973), 279-287. MR 49(1975), 1135.

Myškis, A.D. and Vigant, È.I.:
1955 On the connection of proximity spaces with extensions of topological spaces, Dokl. Akad. Nauk SSSR (N.S.) 103(1955), 969-972. MR 18(1957), 140.

Nadler, S.B., Jr.:
1972 Some results and problems about embedding certain compactifications, Proceedings of the University of Oklahoma Topology Conference, 222-233, Univ. of Oklahoma, Norman, Okla., 1972. MR 50(1975), 2013-2014.

Nadler, S.B., Jr. and Quinn, J.:
1973 Embedding certain compactifications of a half-ray, Fund. Math. 78(1973), 217-225. MR 47(1974), 1678.

Nadler, S.B., Jr., Quinn, J., and Reiter, H.:
1975 Results and problems concerning compactifications, compact subtopologies, and mappings, Fund. Math. 89(1975), 34-44.

Nagata, Jun-iti:
1950 On the uniform topology of bicompactifications, J. Inst. Polytech. Osaka City Univ. Ser. A. Math 1(1950), 28-38. MR 12(1951), 272.

Narici, L., Beckenstein, E., and Bachman, G.:
1974 Some recent developments on repletions and Stone-Čech compactifications of 0-dimensional spaces, TOPO 72 - General Topology and its Applications, 310-321, Lecture Notes in Math., V. 378, Springer-Verlag, Berlin, 1974. MR 50(1975), 2005.

Negrepontis, S.:
1967 Absolute Baire sets, Proc. AMS 18(1967), 691-694. MR 35(1968), 903.

1967 Baire sets in topological spaces, Arch. Math. (Basel) 18(1967), 603-608. MR 36(1968), 669.

1967 A note on the r-compactifications of some discrete sets, Arch. Math. (Basel) 18(1967), 264-266. MR 36(1968), 186.

1968 Extension of continuous functions in βD, Nederl. Akad. Wetensch. Proc. Ser. A 71 = Indag. Math. 30(1968), 393-400. MR 39(1970), 393.

1969 Extension of continuous functions in βD, Contributions to Extension Theory of Topological Structures (Proc. Sympos., Berlin, 1967), 163-169. Deutsch. Verlag Wissensch., Berlin, 1969. MR 40(1970), 1183.

1969 An example on realcompactifications, Arch. Math. (Basel) 20(1969), 162-164. MR 39(1970), 1141.

1969 On the product of F-spaces, Trans. AMS 136(1969), 339-346. MR 38(1969), 500.

Nel, L.D. and Riordan, D.:
1972 Note on a subalgebra of C(X), Canad. Math. Bull. 15(1972), 607-608. MR 47(1974), 1013.

Nemec, A.G.:
1974 Bicompact extensions that do not increase the weight and the dimension, Dokl. Akad. Nauk SSSR 214(1974), 517-519. English Translation: Soviet Math. Dokl. 15(1974), 189-192. MR 49(1975), 2107.

Netrebin, A.G.:
1974 The lattices of closed sets of topological spaces (Russian), Ural. Gos. Univ. Mat. Zap. 8, tetrad'4, 81-84, 135(1974). MR 51(1976), 944-945.

Nielsen, R. and Sloyer, C.:
1970 Ideals of semi-continuous functions and compactifications, Math. Ann. 187(1970), 329-331. MR 42(1971), 1533.

Nillsen, R.:
1969 Compactification of products, Mat. Časopis Sloven. Akad. Vied 19(1969), 316-323. MR 46(1973), 1409.

1975 Discrete orbits in $\beta N \setminus N$, Colloq. Math. 33(1975), 71-81, 160.

Njåstad, O.:
1965 A note on compactification by bounding systems, J. London Math. Soc. 40(1965), 526-532. MR 33(1967), 326.

1966 On Wallman-type compactifications, Math. Z. 91(1966), 267-276. MR 32(1966), 1096.

1975 Multiple points and Wallman compactifications, J. Austral. Math. Soc. 20(1975), 274-280.

Nöbeling, G. and Bauer, H.:
1955 Über die Erweiterungen topologischer Räume, Math. Ann. 130(1955), 20-45. MR 17(1956), 390.

Novák, J.:
1953 Über die bikompakte Hülle einer isolierten abzählbaren Menge, Bericht über die Mathematiker-Tagung in Berlin, Jan., 1953, pp. 280-283. Deutscher Verlag der Wissenschaften, Berlin, 1953. MR 16(1955), 608.

Nowiński, K.:
1972 Closed mappings and the Freudenthal compactification, Fund. Math. 76(1972), 71-83. MR 48(1974), 521.

1974 Extension of closed mappings, Fund. Math. 85(1974), 9-17. MR 51(1976), 582.

Okuyama, A.:
1971 A characterization of a space with countable infinity, Proc. AMS 28(1971), 595-597. MR 43(1972), 505.

Onuchic, N.:
1957 P-spaces and Stone-Čech compactification, Boll. Soc. Math. São Paulo 12(1957), 11-41. MR 23(1962), 407.

Osmatesku, P.K.:
1963 $\omega\alpha$-compactifications, Vestnik Moskov. Univ. Ser. I Mat. 1963, No. 6, 45-54. MR 28(1964), 121.

1964 Generalization of a one-point bicompactification theorem of P.S. Alexandroff, Dokl. Akad. Nauk SSSR 157(1964), 274-275. MR 29(1965), 117.

1965 On continuation of perfect mappings to $\omega\alpha$-extensions, Dokl. Akad. Nauk SSSR 164(1965), 985-988. Translation: Soviet Math. Dokl. 6(1965), 1336-1339. MR 32(1966), 1098-1099.

1969 Proximity on T_1-spaces (Russian) Czech. Math. J. 19(94) (1969), 193-207. MR 40(1970), 363.

1970 Bicompact extensions $\omega_L X$ of T_1-spaces (Russian), Bull. Acad. Polon. Sci. Sér. Sci. Math. Astronom. Phys. 18(1970), 579-587. MR 43(1972), 214-215.

1972 Relative connectedness and perfect compactifications of a T_1-space (Russian), Bull. Acad. Polon. Sci. Sér. Sci. Math. Astronom. Phys. 20(1972), 25-35. MR 46(1973), 1084.

Osmatesku, P.K. and Čoban, M.M.:
1975 Bicompactifications of peripherally bicompact spaces (Russian), Mat. Issled. 10(1975), vyp. 1(35), 276-286, 303-304.

Ovčinnikov, I.S.:
1966 Prime ends of a certain class of space regions, Trudy Tomsk. Gos. Univ. Ser. Meh.-Mat. 189(1966), 96-103. MR 37(1969), 412.

Ovsepjan, S.G.:
1974 Bicompact and H-closed finite point extensions of topological spaces (Russian, Armenian summary), Akad. Nauk Armjan. SSR Dokl. 59(1974), 23-28.

1975 The lattice structure of the set of semi-regular H-closed extensions of topological spaces (Russian, Armenian summary), Akad. Nauk Armjan. SSR Dokl. 61(1975), 137-140.

Panteleev, V.P.:
1966 Hausdorff extensions of topological spaces, Trudy Naučn. Ob" ed. Prepodav. Fiz.-Mat. Fak. Ped. Inst. Dal'n. Vostok. 7(1966), 68-76. MR 35(1968), 428.

Papy, G.:
1953 Sur les compactifications d'Alexandroff, Acad. Roy. Belgique. Bull. Cl. Sci. (5) 39(1953), 937-941. MR 15(1954), 546.

Pareek, C.M.:
1971 Characterizations of p-spaces, Canad. Math. Bull. 14(1971), 459-460. MR 47(1974), 1013.

Park, Y.L.:
1969 On the projective cover of the Stone-Čech compactification of a completely regular Hausdorff space, Canad. Math. Bull. 12(1969), 327-331. MR 40(1970), 906.

Parovičenko, I.I.:
1963 On a universal bicompactum of weight ℵ, Dokl. Akad. Nauk SSSR 150(1963), 36-39. MR 27(1964), 153-154.

1970 A remark on an article by S. Iliadis and S. Fomin, Uspehi Mat. Nauk 25(1970), No. 6 (156), 245-246. MR 45(1973), 489.

Pasynkov, B.A.:
1967 Universal bicompacta and metric spaces of given dimension (Russian), Fund. Math. 60(1967), 285-308. MR 36(1968), 1362-1363.

Pearson, B.J.:
1973 Dendritic compactifications of certain dendritic spaces, Pacific J. Math. 47(1973), 229-232. MR 48(1974), 874.

Pelletier, D.H.:
1976 A note on defining the Rudin-Keisler ordering of ultrafilters, Notre Dame J. Formal Logic 17(1976), 284-286.

Peregudov, S.A.:
1975 Certain properties of bicompact extensions (Russian), Mat. Zametki 17(1975), 467-473.

Petrisor, P.A.:
1973 Sur un type de compactification, Matematica (Cluj) 15(38) (1973), 263-287.

Pfeifer, G.L.:
1969 The Stone-Čech compactification of an irreducibly connected space, Proc. AMS 20(1969), 531-532. MR 38(1969), 686.

Piacun, N. and Su, L.P.:
1973 Wallman compactifications on E-completely regular spaces, Pacific J. Math. 45(1973), 321-326. MR 47(1974), 1338.

Pickert, G.:
1959 Erweiterungen eines topologischen Räumes, Arch. Math. 10(1959), 155-161. MR 21(1960), 813-814.

Plank, D.:
1969 On a class of subalgebras of C(X) with applications to $\beta X \setminus X$, Fund. Math. 64(1969), 41-54. MR 39(1970), 1141.

Poljakov, V.Z.:
1969 On some proximity properties determined only by the topology of the compactifications, Contributions to Extension Theory of Topological Structures (Proc. Sympos., Berlin, 1967), 173-178. Deutsch Verlag Wissensch., Berlin, 1969. MR 40(1970), 1180.

Pondiczery, E.S.:
1944 Power problems in topological spaces, Duke Math. J. 11(1944), 835-837.

Ponomarev, V.I.:
1960 Extensions of many-valued mappings of topological spaces to their compactifications, Mat. Sb. (N.S.) 52(94) (1960), 847-862. MR 22(1961), 2152.

1962 On spaces satisfying countability axioms, Vestnik Moskov. Univ. Ser. I Mat. Meh. 1962, No. 4, 44-50. MR 33(1967), 327.

1963 On the absolute of a topological space, Dokl. Akad. Nauk SSSR 149(1963), 26-29. MR 28(1964), 122-123.

1964 On Wallman's compactification of a topological space, Sibirsk. Mat. Ž. 5(1964), 1333-1342. MR 32(1966), 290-291.

Porter, J. and Thomas, J.:
1969 On H-closed and minimal Hausdorff spaces, Trans. AMS 138(1969), 159-170. MR 38(1969), 1178-1179.

Porter, J.R. and Woods, R.G.:
1972 Nowhere dense subsets of metric spaces with applications to Stone-Čech compactifications, Canad. J. Math. 24(1972), 622-630. MR 48(1974), 526-527.

Pospíšil, B.:
1937 Remark on bicompact spaces, Ann. Math. 38(1937), 845-846.

Purisch, S.:
1973 On the orderability of Stone-Čech compactifications, Proc. AMS 41(1973), 55-56. MR 48(1974), 870-871.

Prodanov, Iv.:
1967 Compact representations of continuous algebraic structures (Russian), General Topology and its Relations to Modern Analysis and Algebra, II (Proc. Second Annual Prague Topological Sympos., 1966), 290-294. Academia, Prague, 1967. MR 38(1969), 686.

Pym, J.S.:
1963 On almost periodic compactifications, Math. Scand. 12(1963), 189-198. MR 29(1965), 118-119.

Raimi, R.A.:
1966 Homeomorphisms and invariant measures for $\beta N \setminus N$, Duke Math. J. 33(1966), 1-12. MR 33(1967), 1118.

Rajagopalan, M.:
1972 $\beta N \setminus N \setminus \{p\}$ is not normal, J. Indian Math. Soc. (N.S.) 36(1972), 173-176. MR 47(1974), 1678.

Rakowski, Z.M.:
1975 On decompositions of compact Hausdorff spaces, Bull. Acad. Polon. Sci. Sér. Sci. Math. Astronom. Phys. 23(1975), 1089-1091.

Rao, B.V.:
1969 Compactification of a totally ordered set, Math. Student 37(1969), 75-80. MR 41(1971), 1698.

Rao, C.J.M.:
1974 On the smallest compactification for convergence spaces, Proc. AMS 44(1974), 225-230. MR 48(1974), 1659.

1975 On the largest Hausdorff compactification for convergence spaces, Bull. Austral. Math. Soc. 12(1975), 73-79. MR 50(1975), 2003.

Rayburn, M.C.:
1969 On the lattice of compactifications and the lattice of topologies, Thesis, University of Kentucky, 1969.

1973 A characterization of realcompact extensions, Proc. AMS 40(1973), 331-332. MR 48(1974), 208.

1973 On the Stoilow-Kerékjártó compactification, J. London Math. Soc. (2) 6(1973), 193-196. MR 48(1974), 208.

1973 On Hausdorff compactifications, Pacific J. Math. 44(1973), 707-714. MR 47(1974), 1013-1014.

1975 A maximal realcompactification with 0-dimensional outgrowth, Proc. AMS 51(1975), 441-447. MR 51(1976), 1605.

Reichaw-Reichbach, M.:
1963 On compactification of metric spaces, Israel J. Math. 1(1963), 61-74. MR 28(1964), 878.

Richardson, G.D.
1970 A Stone-Čech compactification for limit spaces, Proc. AMS 25(1970), 403-404. MR 41(1971), 179.

Richardson, G.D. and Kent, D.C.:
1972 Regular compactifications of convergence spaces, Proc. AMS 31(1972), 571-573. MR 44(1972), 616.

Rinow, W.:
1964 Perfekte lokal zusammenhängende Kompaktifizierungen und Primendtheorie, Math. Z. 84(1964), 294-304. MR 29(1965), 323.

1969 Die Freudenthalsche Trennungsrelation und Berührungsstrukturen, Contributions to Extension Theory of Topological Structures (Proc. Sympos., Berlin, 1967), 189-192. Deutsch. Verlag Wissensch., Berlin, 1969. MR 40(1970), 363.

Robison, S.M.:
1969 Some properties of $\beta X \setminus X$ for complete spaces, Fund. Math. 64(1969), 335-340. MR 39(1970), 1141-1142.

Rogers, J.W., Jr.:
1969 On compactifications with continua as remainders, Notices AMS 16(1969), 1045, abstract 669-9.

1971 On compactifications with continua as remainders, Fund. Math. 70(1971), 7-11. MR 44(1972), 192.

Rudd, D.:
1975 A note on zero-sets in the Stone-Čech compactification, Bull. Austral. Math. Soc. 12(1975), 227-230. MR 51(1976), 1602.

Rudin, M.E.:
1966 Types of Ultrafilters, Topology Seminar (Wisconsin, 1965), pp. 147-151, Math. Studies, No. 60, Princeton University Press, Princeton, 1966. MR 35(1968), 1354.

1970 Composants and βN, Proc. Washington State Univ. Conf. on General Topology (Pullman, Wash., 1970), pp. 117-119. MR 42(1971), 190.

1971 Partial orders on the types in βN, Trans. AMS 155(1971), 353-362. MR 42(1971), 1535.

1975 Lectures on Set Theoretic Topology, American Mathematical Society, Providence, R.I., 1975, iv + 76 pp. MR 51(1976), 579.

Rudin, W.:
1956 Homogeneity problems in the theory of Čech compactifications, Duke Math. J. 23(1956), 409-419. Note of Correction, Duke Math. J. 23(1956), 633. MR 18(1957), 324.

Rudolf, L.:
1967 On compactification of T_o-spaces, Colloq. Math. 17(1967), 41-50. MR 35(1968), 661.

Ryll-Nardzewski, C. and Telgársky, R.:
1970 On the scattered compactification, Bull. Acad. Polon. Sci. Sér. Sci. Math. Astronom. Phys. 18(1970), 233-234. MR 41(1971), 1400.

Saegrove, M.J.:
1973 Pairwise complete regularity and compactification in bitopological spaces, J. London Math. Soc. (2) 7(1973), 286-290. MR 49(1975), 2109.

Salbany, S.:
1974 On compact* spaces and compactifications, Proc. AMS 45(1974), 274-280. MR 50(1975), 1152-1153.

Salbany, S. and Brümmer, G.C.L.:
1971 Pathology of upper Stone-Čech compactifications, Amer. Math. Monthly 78(1971), 187-188. MR 43(1972), 215.

Samuel, Pierre:
1948 Ultrafilters and compactification of uniform spaces, Trans. AMS 64(1948), 100-132. MR 10(1949), 54.

Šapiro, L.B.:
1974 Three examples in the theory of bicompact extensions (Russian), Dokl. Akad. Nauk SSSR 217(1974), 774-776.

1974 Bicompact extensions of Wallman type, Vestnik Moskov. Univ. Ser. I Mat. Meh. 29(1974), No. 5, 19-23. English Translation: Moscow Univ. Math. Bull. 29(1974), No. 5, 16-19. MR 50(1975), 2009.

1974 A reduction of the fundamental problem of bicompact extensions of Wallman type (Russian), Dokl. Akad. Nauk SSSR 217(1974), 38-41. Translation: Soviet Math Dokl. 15(1974), 1020-1023.

Šapirovskii, B.E.:
1975 The imbedding of extremally disconnected spaces in bicompacta. b-points and weight of pointwise normal spaces (Russian), Dokl. Akad. Nauk SSSR 223(1975), 1083-1086.

Ščepin, E.:
1970 The spaces that are close to normal ones and their bicompact extensions, Dokl. Akad. Nauk SSSR 191(1970), 295-297. Translation: Soviet Math. Dokl. 11(1970), 373-375. MR 41(1971), 826-827.

1972 The bicompact Ponomarev-Zaicev extension and the so-called spectral parasite (Russian), Mat. Sb. (N.S.) 88(130) (1972), 316-326. MR 46(1973), 451.

Schoenfield, A.H.:
1974 Continuous surjections from Cantor sets to compact metric spaces, Proc. AMS 46(1974), 141-142.

Schurle, A.W.:
1967 Compactification of strongly countable dimensional spaces, Bull. AMS 73(1967), 909-912. MR 36(1968), 434.

1969 Compactifications of strongly countable-dimensional spaces, Trans. AMS 136(1969), 25-32. MR 38(1969), 502.

Šedivá, V.:
1959 Some examples of topological spaces in which the axiom F does not hold, Časopis Pěst. Mat. 84(1959), 461-466. MR 23(1962), 238.

Semadeni, Z.:
1959 Sur les ensembles clairsemés, Rozprawy Mat. 19(1959), 39 pp. MR 21(1960), 1227.

1964 Periods of measurable functions and the Stone-Čech compactification, Amer. Math. Monthly 71(1964), 891-893. MR 30(1965), 1225.

1971 Banach Spaces of Continuous Functions, PWN-Polish Scientific Publishers, Warsaw, 1971. MR 45(1973), 1054-1056.

Shanin, N.A.:
1943 On special extensions of topological spaces, C.R.(Doklady) Acad. Sci. URSS (N.S.) 38(1943), 6-9. MR 5(1944), 45-46.

1943 On separation in topological spaces, C.R. (Doklady) Acad. Sci. URSS (N.S.) 38(1943), 110-113. MR 5(1944), 46.

1943 On the theory of bicompact extensions of topological spaces, C.R. (Doklady) Acad. Sci. URSS (N.S.) 38(1943), 154-156. MR 5(1944), 46.

Shiraki, M.:

1968 Compactification of T_o-spaces, Rep. Fac. Sci. Kagoshima Univ. 1(1968), 9-12, MR 39(1970), 1365-1366.

Shirota, T.:

1950 On systems of structures of a completely regular space, Osaka Math. J. 41(1950), 131-143. MR 13(1952), 764.

Sierpiński, W.:

1938 Fonctions additives non complètement additives et fonctions non measurables, Fund. Math. 30(1938), 96-99.

1965 Cardinal and Ordinal Numbers, PWN-Polish Scientific Publishers, Warsaw, 1965.

Silva, C.:

1969 Liaison spaces and compact extensions, Univ. Habana 33(1969), 111-117. MR 42(1971), 188.

Simon, B.:

1969 Some pictorial compactifications of the real line, Amer. Math. Monthly 76(1969), 536-538. MR 39(1970), 1366.

Skljarenko, E.G.:

1958 On the embedding of normal spaces into bicompacta of the same weight and dimension, Dokl. Akad. Nauk SSSR 123(1958), 36-39. MR 20(1959), 1006.

1958 Bicompact extensions of semi-bicompact spaces, Dokl. Akad. Nauk SSSR 120(1958), 1200-1203. MR 20(1959), 706.

1961 Perfect bicompact extensions, Dokl. Akad. Nauk SSSR 137(1961), 39-41. Translation: Soviet Math. Dokl. 2, 238-240. MR 22(1961), 2152.

1961 On the extension of homeomorphisms, Dokl. Akad. Nauk SSSR 141(1961), 1045-1047. MR 25(1963), 499.

1962 Some questions in the theory of bicompactifications, Izv. Akad. Nauk SSSR Ser. Mat. 26(1962), 427-452. MR 26(1963), 141.

1962 On perfect bicompact extensions, Dokl. Akad. Nauk SSSR 146(1962), 1031-1034. MR 25(1963), 1073.

1964 Some applications of the theory of sheaves in general topology, Uspehi Mat. Nauk 19(1964), No. 6(120), 47-70. MR 30(1965), 294.

1969 Uniform edging structures, Dokl. Akad. Nauk SSSR 186(1969), 39-42. Translation: Soviet Math. Dokl. 10(1969), 559-562. MR 40(1970), 163-164.

Skula, L.:

1969 Ordered set of classes of compactifications, Czech. Math. J. 19(94) (1969), 42-59. MR 38(1969), 1179.

Slowikowski, W. and Zawadowski, W.:
1955 A generalization of maximal ideals method of Stone and Gelfand, Fund. Math. 42(1955), 215-231. MR 18(1957), 223.

Smirnov, Ju. M.:
1951 On coverings of topological spaces, Moskov. Gos. Univ. Učenye Zapiski 148, Matematika 4(1951), 204-215. MR 14(1953), 303.

1951 Some relations in the theory of dimensions, Mat. Sbornik N.S. 29(71) (1951), 157-172. MR 13(1952), 372.

1952 On proximity spaces, Mat. Sb. (N.S.) 31(73) (1952), 543-574. Translation: AMS Transl. (2) 38(1964), 5-35. MR 14(1953), 1107.

1952 On proximity spaces in the sense of V.A. Efremovič, Dokl. Akad. Nauk SSSR 84(1952), 895-898. MR 14(1953), 1107.

1958 A completely regular non-semibicompact space with a zerodimensional Čech complement, Dokl. Akad. Nauk SSSR 120(1958), 1204-1206. MR 20(1959), 706.

1965 Über die Dimension der Adjunkten bei Kompaktifizierungen, Monats. Deutsch. Akad. Wiss. Berlin 7(1965), 230-232, 750-753. MR 33(1967), 560.

1966 Dimension of increments of proximity spaces and of topological spaces, Dokl. Akad. Nauk SSSR 168(1966), 528-531. Translation: Soviet Math. Dokl. 7(1966), 688-691, supp. note, ibid. 7(1966), 691-692. MR 35(1968), 903.

1966 On the dimension of remainders in bicompact extensions of proximity and topological spaces, II, Mat. Sb. (N.S.) 71(113) (1966), 454-482. MR 35(1968), 903-904.

1966 On the dimension of remainders in bicompact extensions of proximity and topological spaces, Mat. Sb. (N.S.) 69(111) (1966), 141-160. Translation: AMS Transl. (2) 84(1969), 197-217. MR 33(1967), 1113.

1967 Proximity and construction of compactifications with given properties, General Topology and its Relations to Modern Analysis and Algebra, II (Proc. Sympos, Prague, 1966), pp. 332-340, Academia, Prague, 1967.

1968 Contractions to bicompacta and their connection with retractions and bicompact extensions (Russian), Fund. Math. 63(1968), 199-211. MR 38(1969), 686.

Snyder, A.K.:
1969 The Čech compactification and regular matrix summability, Duke Math. J. 36(1969), 245-252. MR 40(1970), 365.

Solomon, R.C.:
1973 A type of βN with $\aleph_0$ relative types, Fund. Math. 69(1973), 209-212. MR 47(1974), 1445.

Šostak, A.P.:
1974 E-compact extensions of topological spaces (Russian), Funkcional. Anal. i Priložen. 8(1974), 62-68.

Steiner, E.F.:
1966 Normal families and completely regular spaces, Duke Math. J. 33(1966), 743-745. MR 33(1967), 1391-1392.

1967 Wallman spaces and compactifications, Fund. Math. 61(1967/68), 295-304. MR 36(1968), 1140.

Steiner, A.K. and Steiner, E.F.:
1968 Compactifications as closures of graphs, Fund. Math. 63(1968), 221-223. MR 38(1969), 1179.

1968 Products of compact metric spaces are regular Wallman, Nederl. Akad. Wetensch. Proc. Ser. A 71=Indag. Math 30(1968), 428-430. MR 40(1970), 162.

1969 On countable multiple point compactifications, Fund. Math. 65(1969), 133-137. MR 40(1970), 162.

1970 Nest generated intersection rings in Tychonoff spaces, Trans. AMS 148(1970), 589-601. MR 41(1971), 1400.

1970 Graph closures and metric compactifications of N, Proc. AMS 25(1970), 593-597. MR 41(1971), 1696.

1971 Relative types of points in $\beta N \setminus N$, Trans. AMS 160(1971), 279-286. MR 49(1975), 258.

1972 Binding spaces: A unified completion and extension theory, Fund. Math. 76(1972), 43-61. MR 46(1973), 1409.

1976 On the reduction of the Wallman compactification problem to discrete spaces, Gen. Topology and Appl., to appear.

Stone, A.H.:
1959 Cardinals of closed sets, Mathematika 6(1959), 99-107. MR 22(1961), 844.

Stone, M.H.:
1937 Applications of the general theory of Boolean Rings to general topology, Trans. AMS 41(1937), 375-481.

1948 On the compactification of topological spaces, Ann. Soc. Polon. Math. 21(1948), 153-160. MR 10(1949), 137.

Su, L.P.:
1974 Wallman type compactifications on 0-dimensional spaces, Proc. AMS 43(1974), 455-460.

1976 Round subsets of Wallman-type compactifications, J. Austral. Math. Soc. Ser. A 21(1976), 224-233.

1976 Wallman type compactifications on 0-**dimensional** spaces, Proc. AMS, to appear.

Sultan, A.:
1975 Lattice realcompactifications, Ann. Math. Pura Appl. (4) 106(1975), 293-303.

Suvorov, G.D.:
1953 Prime ends of a sequence of plane regions converging to a nucleus, Mat. Sbornik N.S. 33(75), (1953), 73-100. MR 15(1954), 244.

Taĭmanov, A.D.:
1952 On extension of continuous mappings of topological spaces (Russian), Mat. Sb. 31(1952), 459-463. MR 14(1953), 395.

1961 Extension of monotone mappings to monotone mappings of compact spaces, Dokl. Akad. Nauk SSSR 135(1960), 23-25. Translated in Soviet Math. Dokl. 1(1961), 1236-1238. MR 23(1962), 240.

Takeuti, G. and Zaring, W.M.:
1971 Introduction to Axiomatic Set Theory, Springer-Verlag, New York, Heidelberg, Berlin, 1971.

Tall, F.D.:
1974 P-points in $\beta N \setminus N$, normal nonmetrizable Moore spaces and other problems of Hausdorff, TOPO 72, pp. 501-512. Lecture Notes in Math., V. 378, Springer, Berlin, 1974.

Tamano, H.:
1960 Some properties of the Stone-Čech compactification, J. Math. Soc. Japan 12(1960), 104-117. MR 26(1963), 1314.

1960 A note on the pseudo-compactness of the product of two spaces, Mem. Coll. Sci. Univ. Kyoto Ser. A Math. 33(1960/61), 225-230. MR 22(1961), 1941.

1961 On compactifications, J. Math. Kyoto Univ. 1(1961/62), 161-193. MR 25(1963), 1072-1073.

1969 The role of compactifications in the theory of Tychonoff spaces, Contributions to Extension **Theory** of Topological Structures (Proc. Sympos. Berlin, 1967) pp. 219-220. Deutsch Verlag Wissensch., Berlin, 1969.

Tanre, D.:
1971 Compactification de Stone-Čech et triple associé, Esquisses Math., No. 16, 29 pp., Fac. Sci. Univ. Paris VII, Paris, 1971.

Terada, T.:
1974 On the dimension of the remainder of Stone-Čech compactifications, Sci. Rep. Tokyo Kyoiku Daigaku Sect. A 12(1974), 218-221.

Terasaka, H.:
1952 On the Cartesian product of compact spaces, Osaka Math. J. 4(1952), 11-15. MR 14(1953), 489.

Thrivikraman, T.:
1971 On localcompactifications of Tychonoff spaces, Kyungpook Math. J. 11(1971), 151-153. MR 46(1973), 1409-1410.

1972 On Hausdorff quotients and Magill's theorem, Monatsh. Math. 76(1972), 345-355. MR 47(1974), 1013.

1972 On the lattices of compactifications, J. London Math. Soc. (2) 4(1972), 711-717. MR 45(1973), 1103.

1972 On compactifications of Tychonoff spaces, Yokohama Math. J. 20(1972), 99-105. MR 46(1973), 768.

1974 Lattice structure of general topological extensions, Kyungpook Math. J. 14(1974), 71-77.

Thron, W.J.:
1966 Topological Structures, Holt, Rinehart, and Winston, New York-Toronto, 1966. MR 34(1967), 130-131.

Todd, C.:
1971 On the compactification of products, Canad. Math. Bull. 14(1971), 591-592. MR 46(1973), 1728.

Tong, H.:
1949 On some problems of Čech, Ann. of Math. (2) 50(1949), 154-157. MR 10(1949), 315.

1970 Solutions of problems of P.S. Alexandroff on extensions of topological spaces, Ann. Mat. Pura Appl. (4) 86(1970), 47-51. MR 43(1972), 215.

Tukey, J.W.:
1940 Convergence and Uniformity in Topology, Annals of Math. Studies, No. 2, Princeton, 1940.

Tychonoff, A.:
1930 Über die topologische Erweiterung von Räumen, Math. Ann. 102(1930), 544-561.

Tzung, F.-C.:
1976 Sufficient conditions for the set of Hausdorff compactifications to be a lattice, Thesis, N.C. State University, 1976.

Ul'janov, V.M.:
1974 Bicompact extensions with the first axiom of countability that do not increase weight and dimension (Russian), Dokl. Akad. Nauk SSSR 217(1974), 1263-1265.

1974 Bicompact extensions with the first axiom of countability, and continuous mappings, Mat. Zametki 15(1974), 491-499. English Translation: Math. Notes 15(1974), 287-291. MR 51(1976), 1605.

1975 Examples of finally compact spaces that do not have bicompact extensions of countable nature (Russian), Dokl. Akad. Nauk SSSR 220(1975), 1282-1285.

1975 Bicompact extensions of countable nature, and absolutes (Russian), Mat. Sb. (N.S.) 98(140) (1975), 223-254, 334.

Ünlü, Y.
1976 Equivalence of Wallman compactifications for locally compact Hausdorff spaces and discrete spaces, Proc. AMS, to appear.

Urysohn, P.:
1925 Über die Mächtigkeit der zusammenhängenden Mengen, Math. Ann. 94(1925), 262-295.

Valiev, V.A.:
1971 The spectral dimension of compact extensions and of products of topological spaces, Dokl. Akad. Nauk SSSR 200(1971), 262-265. Translation: Soviet Math. Dokl. 12(1971), 1358-1362. MR 45(1973), 207.

1972 The spectral dimension of bicompact extensions, Vestnik Moskov. Univ. Ser. I. Mat. Meh. 27(1972), 34-42. Translation: Moscow Univ. Math. Bull. 27(1972), 27-33. MR 48(1974), 1248-1249.

van der Slot, J.:
1968 Some properties Related to Compactness, Mathematical Centre Tracts, 19, Mathematisch Centrum, Amsterdam, 1968. MR 40(1970), 163.

1966 Universal topological properties, Math. Centrum Amsterdam Afd. Zuivere Wisk. 1966, ZW-011, 9 pp. MR 39(1970), 632.

Vedenisov, N.B.:
1948 Bicompact spaces, Uspehi Matem. Nauk (N.S.) 3(1948), 67-79. MR 10(1949), 137.

Venkataraman, M., Rajagopalan, M., and Soundararajan, T.:
1972 Orderable topological spaces, General Topology and Appl. 2(1972), 1-10, MR 45(1973), 1406.

Visliseni, J.:
1971 Kompaktifizierungen mittels lokal zusammenhängender uniformer Stukturen, Math. Nach. 48(1971), 153-177. MR 44(1972), 1080-1081.

Visliseni, Ju. and Flaksmaier, Ju.:
1965 Power and construction of the structure of all compact extensions of a completely regular space, Dokl. Akad. Nauk SSSR 165(1965), 258-260. Translation: Soviet Math. Dokl. 6(1965), 1423-1425. MR 32(1966), 1432.

Wagner, F.J.:
1957 Notes on compactifications. I, II, Nederl. Akad. Wetensch. Proc. Ser. A 60 = Indag. Math. 19(1957), 171-176, 177-181. MR 19(1958), 436.

1964 Normal base compactifications, Nederl. Akad. Wetensch. Proc. Ser. A 67 = Indag. Math. 26(1964), 78-83. MR 28(1964), 498.

Walker, R.C.:
1974 The Stone-Čech compactification, Springer-Verlag, New York, Heidelberg, Berlin, 1974.

Wallace, A.D.:
1951 Extensional invariance, Trans. AMS 70(1951), 97-102. MR 12(1951), 845.

Wallman, H.:
1938 Lattices and topological spaces, Ann. of Math (2) 39(1938), 112-126.

Warren, N.M.:
1970 Extending continuous functions in Stone-Čech compactifications of discrete spaces and in zero-dimensional spaces, Thesis, University of Wisconsin, 1970.

1972 Properties of Stone-Čech compactifications of discrete spaces, Proc. AMS 33(1972), 599-606. MR 45(1973), 205.

Wasilewski, J.S.:
1974 Compactifications, Canad. J. Math. 26(1974), 365-371.

Weier, J.:
1957 Über einen Erweiterungssatz, Monatsh. Math. 61(1957), 51-53. MR 18(1957), 813.

Willard, S.:
1966 Absolute Borel sets in their Stone-Čech compactifications, Fund. Math. 58(1966), 323-333. MR 33(1967), 828.

1969 Embedding metric absolute Borel sets in completely regular spaces, Colloq. Math. 20(1969), 83-88. MR 39(1970), 632.

1970 General Topology, Addison-Wesley, Reading, Mass., 1970. MR 41(1971), 1689-1690.

Weir, M.D.:
1975 Hewitt-Nachbin Spaces, Mathematics Studies, No. 17, North-Holland Publishing Co.; Amsterdam, Oxford; American Elsevier Publishing Co., Inc.: New York, 1975, 270 pp.

Wenjen, C.:
1974 On Fan-Gottesman compactification and P.S. Alexandroff's problems, J. Math. Anal. Appl. 47(1974), 269-283. MR 51(1976), 585-586.

Woods, R.G.:
1968 Certain properties of $\beta X \setminus X$ for σ-compact X, Thesis, McGill University, 1968.

1971 A Boolean algebra of regular closed subsets of $\beta X \setminus X$, Trans. AMS 154(1971), 23-36. MR 42(1971), 951.

1971 Some $\aleph_0$-bounded subsets of Stone-Čech compactifications, Israel J. Math. 9(1971), 250-256. MR 43(1972), 738.

1971 Co-absolutes of remainders of Stone-Čech compactifications, Pacific J. Math. 37(1971), 545-560. Correction: Pacific J. Math. 39(1971), 828. MR 46(1973), 1085.

1971 Homeomorphic sets of remote points, Canad. J. Math. 23(1971), 495-502. MR 43(1972), 1249.

1972 Ideals of pseudocompact regular closed sets and absolutes of Hewitt realcompactifications, General Topology and Appl. 2(1972), 315-331. MR 47(1974), 1339.

1972 On the local connectedness of $\beta X \setminus X$, Canad. Math. Bull. 15(1972), 591-594. MR 47(1974), 449.

1974 A Tychonoff almost realcompactification, Proc. AMS 43(1974), 200-208. MR 48(1974), 1663.

1974 Zero-dimensional compactifications of locally compact spaces, Canad. J. Math. 26(1974), 920-930. MR 50(1975), 444.

1975 Topological extension properties, Trans. AMS 210(1975), 365-385. MR 51(1976), 1605.

Wulbert, D.E.:

1969 Locally connected Stone-Čech compactifications, Contributions to Extension Theory of Topological Structures (Proc. Sympos., Berlin, 1967), 247-248. Deutsch. Verlag Wissensch., Berlin, 1969. MR 39(1970), 1366.

1969 A characterization of C(X) for locally connected X, Proc. AMS 21(1969), 269-272. MR 39(1970), 361.

Wulfsohn, A.:

1972 A compactification due to Fell, Canad. Math. Bull. 15(1972), 145-146. MR 48(1974), 525.

Zaǐcev, V.:

1966 Projection spectra and bicompact extensions, Dokl. Akad. Nauk SSSR 171(1966), 521-524. Translation: Soviet Math. Dokl. 7(1966), 1504-1507. MR 34(1967), 1536.

1968 Some classes of topological spaces and their bicompact extensions, Dokl. Akad. Nauk SSSR 178(1968), 778-779. Translation: Soviet Math. Dokl. 9(1968), 192-193. MR 37(1969), 410.

1968 Semiregular and Hausdorff bicompact extensions, Dokl. Akad. Nauk SSSR 182(1968), 27-30. Translation: Soviet Math. Dokl. 9(1968), 1088-1092. MR 38(1969), 502.

1969 Finite spectra of topological spaces and their limit spaces, Math. Ann. 179(1969), 153-174. MR 39(1970), 393.

1970 The bicompactness and completeness of topological spaces in connection with the theory of absolutes, Dokl. Akad. Nauk SSSR 195(1970), 1014-1017. Translation: Soviet Math. Dokl. 11(1970), 1595-1599. MR 45(1973), 204.

Zambahidze, L.G.:
1970 Certain properties of spaces with bicompact complements of finite order, Dokl. Akad. Nauk SSSR 191(1970), 263-266. Translation: Soviet Math. Dokl. 11(1970), 335-338. MR 41(1971), 1399-1400.

1973 A certain dimension-like function, and its connection with the inductive dimensions and bicompactification theory (Russian), Proc. Inter. Symp. on Top. and its Appl., pp. 242-248. Savez Društava Mat. Fiz. i Astronom., Belgrade, 1973.

Zame, A.:
1969 A note on Wallman spaces, Proc. AMS 22(1969), 141-144. MR 39(1970), 874.

Zarelua, A.:
1962 Equality of dimension and bicompact extensions, Dokl. Akad. Nauk SSSR 144(1962), 713-716. MR 26(1963), 1053.

1964 The continuation of mappings to extensions with certain properties, Sibirsk. Mat. Z. 5(1964), 532-548. MR 30(1965), 478.

1969 The method of the theory of rings of functions in the construction of bicompact extensions (Russian; English summary), Contributions to Extension Theory of Topological Structures (Proc. Sympos., Berlin, 1967), 249-256. Deutsch. Verlag Wissensch., Berlin, 1969. MR 40(1970), 365.

Zenor, P.:
1970 Extending completely regular spaces with inverse limits, Glasnik Mat. Ser. III 5(25), (1970), 157-162. MR 43(1972), 215.

1970 Extending spaces with projective spectra, Glasnik Mat. Ser. III 5(25) (1970), 335-342. MR 45(1973), 798-799.

SYMBOL INDEX

AUTHOR INDEX

SUBJECT INDEX